CONVERSATIONS ON
COGNITIVE CULTURAL STUDIES

"In *Conversations on Cognitive Cultural Studies: Literature, Language, and Aesthetics*, Frederick Luis Aldama and Patrick Colm Hogan illuminate the myriad implications of cognitive studies for the study of literature. Scrutinizing issues ranging from subjectivity to aesthetics to history and politics, they seek ever greater precision in our methods and assumptions. This lively critical dialogue is sure to spark much-needed thought and discussion."

—Sue J. Kim, author of *On Anger: Race, Cognition, Narrative*

"*Conversations on Cognitive Cultural Studies* gives a strong sense of cognitive studies in their present state of organization, as represented by two of its most respected and best informed spokesmen, who are seen grappling with basic and, in some cases, still unresolved problems in aesthetics."

—Irving Massey, professor emeritus of English and comparative literature, University at Buffalo

CONVERSATIONS ON COGNITIVE CULTURAL STUDIES

LITERATURE, LANGUAGE, AND AESTHETICS

Frederick Luis Aldama
AND Patrick Colm Hogan

THE OHIO STATE UNIVERSITY PRESS · COLUMBUS

Copyright © 2014 by The Ohio State University.
All rights reserved.

Library of Congress Cataloging-in-Publication Data
Aldama, Frederick Luis, author.
  Conversations on cognitive cultural studies : literature, language, and aesthetics / Frederick Luis Aldama and Patrick Colm Hogan.
    pages cm
  Includes bibliographical references and index.
  ISBN-13: 978-0-8142-1243-1 (cloth : alk. paper)
  ISBN-10: 0-8142-1243-3 (cloth : alk. paper)
  ISBN-13: 978-0-8142-9346-1 (cd-rom)
  ISBN-10: 0-8142-9346-8 (cd-rom)
  1. Psychology and literature. 2. Cognitive science—Philosophy. 3. Narration (Rhetoric) I. Hogan, Patrick Colm, author. II. Title.
  PN56.P93A43 2014
  801'.92—dc23
                         2013028050

Cover design by Laurence J. Nozik
Text design by Juliet Williams
Type set in Adobe Minion Pro

♾ The paper used in this publication meets the minimum requirements of the American National Standard for Information Sciences—Permanence of Paper for Printed Library Materials. ANSI Z39.48-1992.

9 8 7 6 5 4 3 2 1

CONTENTS

Acknowledgments	vi
Prologue	vii
1 Puzzling Out the Self	1
2 Verbal Art and Language Science	32
3 On Matters of Narrative Fiction	79
4 A Scientific Approach to Aesthetics	116
5 Situating History, Culture, Politics, and Ethics in Literary Studies	149
Works Cited and Suggestions for Further Reading	187
Index	197

ACKNOWLEDGMENTS

An earlier version of chapter 1 appeared as "Puzzling Out the Self: Some Initial Reflections," *English Language Notes* 49.2 (2011): 139–60. We are grateful to the editors, Julie Carr and John-Michael Rivera, for permission to reprint. Herbert Lindenberger, Javier Gutiérrez-Rexach, Irving Massey, Robyn Warhol, and James Phelan provided very helpful comments on earlier versions of this manuscript. Our editor at The Ohio State University Press, Sandy Crooms, was invaluable in the process of planning, evaluating, and producing this book.

PROLOGUE

The following dialogues, which took place at various times between 2010 and 2012, focus on the place of cognitive science in the study of culture, particularly literature. For readers unfamiliar with the field, cognitive science is a recently developed approach to the study of the human mind. It integrates a range of disciplines including anthropology, cognitive psychology, computer science, evolutionary biology, linguistics, and others, with empirical studies of human and nonhuman mental processing. Beyond biological and social sciences, the contributing disciplines include art history, musicology, and, most important for our purposes, literary criticism and theory. In each case, the traditional discipline (e.g., literary study) is enhanced by the contributions of general cognitive scientific theory and research while at the same time the general theory is itself advanced and refined by research in the disciplines, again including literary study. Cutting across the traditional academic disciplines, cognitive science also comprises a number of subfields, including cognitive neuroscience, which focuses on brain research; social cognition, which treats the interaction of human minds in, for example, shared tasks; and affective science, which examines human emotions or motivation systems. These subfields in turn combine with one another and with traditional disciplines to produce further areas of research, such as social neuroscience (the study of the operation of the brain in social interaction), neurolinguistics (the study of the neurological processes of language and speech), and so on.

Thus cognitive science is, in part, a research program shared across a range of disciplines and subfields. But that alone would not give the field much coherence. It is, in addition, defined by a set of theoretical principles about "cognitive architecture." Cognitive architecture comprises structures (the enduring organizational categories of the mind), processes (the operations that occur within the structures), and contents (the objects on which processes operate). For example, most cognitive scientists would agree that there are different types of memory, thus different memory structures. These include episodic memory (concerning events that one experienced), skills memory (concerning how to do things, such as ride a bicycle), semantic memory (concerning meanings and facts), working memory (what we keep in mind at any one time), and so on. These involve processes, such as memory storage and retrieval, as well as contents (the memories themselves). The example of memory may seem to suggest that cognitive architecture is isolated and, so to speak, inward-oriented. However, cognitive architecture is at the basis of social interaction as well. For example, one important component of social cognition is "Theory of Mind" (ToM). ToM is simply the means by which we come to understand what other people are thinking or feeling.

The cases of memory and ToM are instances of what is sometimes called "phenomenological" or "mentalistic" architecture. This sort of architecture is to some extent intuitively comprehensible, since it is based on our ordinary understanding of various mental structures and processes. We all know intuitively what "memory" is, even if we do not spontaneously organize it in the way that is done by cognitive scientists—and certainly not with the precision and empirical guidance of cognitive scientists. There is also a "neurocognitive" architecture correlated with this mentalistic architecture. The neurocognitive architecture comprises the structures and processes of the human brain—for instance, the regions of functional neuroanatomy. Thus, in speaking of neurocognitive architecture, we might refer to the hippocampus. It is a crucial part of current cognitive science that the phenomenological architectures are continually revised by reference to the neurocognitive architectures while the neurocognitive architectures are understood by reference to the phenomenological architectures—as when we speak of the operation of the hippocampus in episodic memory.

Beyond these two basic architectures, we may also distinguish different sorts of cognitive theory. The basic architectures are fairly general. The theories tend to be more particular and are often developed in relation to specific cognitive domains (e.g., language). For example, different linguistic theories accept that there is something along the lines of semantic memory, but they specify and explain this in sometimes very different ways. These differences are often related to the theories' precise relation to basic architectures. Thus in linguistics we may distinguish theories that stress mentalistic architecture from theories that stress neurocognitive architecture. The former find one influential form in Noam Chomsky's work, which is based on sets of rules with variables. The neurocognitive approach appears in a rather abstract form in Connectionism, which posits only networks of interlinked neuron-like units with distinct activation patterns. Cutting across this division is a distinction between theories that rely solely on general architecture and those that draw on a specialized architecture for the domain at issue. For example, both "Generative" linguists, such as Chomsky, and "Cognitive" linguists, such as George Lakoff and Ronald Langacker, stress a mentalistic architecture (though they draw on neurocognitive research in formulating and developing that architecture). However, Generative linguists tend to maintain that the general architecture must be supplemented by principles specific to language, whereas Cognitive linguists try to explain language processes by principles available in the general architecture.[1]

The example of language is particularly apt here for the obvious reason that language is fundamental to literature. Thus understanding language is presumably fundamental to our understanding of literature. A simple, communicative model of literature involves at least four components: 1) a speaker or author, 2) a listener or reader, 3) an utterance or text, and 4) a social and historical context. We have used this basic structure to organize the topics for our dialogues. The first chapter addresses the fundamental element of any communicative act, the component most obviously related to cognitive neuroscience—the person, the self, either as speaker or as hearer. The final chapter turns to the social and historical context. This may seem most problematic for neuroscientific treatment, but it is an area with rich possibilities. The intervening chapters concentrate on the text, though with constant reference to the selves that actualize those texts. Specifically, chapter 2 explores the linguistic aspect of verbal art (including treatment of the linguistic theories briefly mentioned above). Chapter 3 takes up narrative. Finally, chapter 4 turns to aesthetics, the area of research that is most directly concerned with the distinctive features of verbal art. In each of these cases—linguistics, narratology,

---

1. Chomskyan and related approaches are commonly referred to as "Generativist," since they set out to generate the sentences of a given language from a set of (language-specific) principles. The name "Connectionism" derives from that theory's stress on network-defining connections across units. "Cognitive Linguistics" is so called due to its use of general cognitive principles (rather than language-specific principles) to explain linguistic processes. Unfortunately, the word "Cognitive" in "Cognitive Linguistics" can be confusing for readers new to the field. Cognitive Linguistics is only one form of linguistic theory within cognitive science. (We will consider these theories in greater detail in chapter 2.)

and aesthetics—cognitive neuroscience has begun to demonstrate its great value. However, its potential contribution to these fields extends much further.

One purpose of these dialogues is to indicate some of what has been accomplished thus far. But a more important purpose is to suggest what may be accomplished in the future. In a field that is far from settled—indeed, is very much alive with debates and disagreements—we felt that a dialectical approach would be most fruitful for both purposes. Our hope is that, in our discussions, we share enough to speak to one another productively, but differ enough to hint at the breadth of the field and to provoke further thought and debate on these important issues. In short, our hope is that these take up at least the spirit of Socratic dialogues.

PCH (for the authors)

CHAPTER 1

Puzzling Out the Self

FREDERICK LUIS ALDAMA: Our writing on the self is *not* new territory; it is one that has been crisscrossed by all variety of scholarly inquiry for over two millennia, from ancient Greek pre-Socratic philosophy, to eighteenth-century metaphysics, nineteenth-century psychology and anthropology, to twentieth-century poststructuralist antifoundationalist skepticism, to today's empirical-based cognitive and neuroscience research. With its long tradition of rich and learned discussion and debate that encompasses all variety of disciplines, any stab at this puzzling out of the self, which will always be constrained by the physical limitations of printed matter, is bound to bewilder.

PATRICK COLM HOGAN: The history of profound (and humbling) reflections on the self is not confined to the West either. Take the Vedāntic tradition of ancient India. In classical Advaita (nondual) Vedānta, the self is a purely spiritual entity that mistakenly identifies itself with the (illusory) material world. In Sāṁkhya, it is a spiritual entity that observes real but distinct material processes that are part of a material world—processes that are simultaneously mental and bodily. Note that, in this tradition, the "ātman" or self is distinct from mind, but mind is composed of the same constituents as body (specifically, the "guṇas" or "strands"). In keeping with this, a third, materialist school accepted only the existence of "prakṛti" or "nature," the world of the *guṇas*, rejecting any distinct spiritual world. Finally, absolute monism accepted both a spiritual world or *ātman* and a material world of *prakṛti* (itself both mind and body, comprising the same component *guṇas*). However, it viewed them as two aspects of a single ("monistic") reality.

To complicate matters even further, there is an additional division of schools into those that accept distinctness of selves and those that do not. In Patañjali's *Yoga Sūtras*, for example, the "observing self" is pretty clearly individual. Indeed, it reaches its highest form in absolute solitude. However, in most Vedāntic schools (schools deriving from the "end" or "conclusion" of the Vedas in the Upaniṣads), the ultimate form of the self, the *ātman* (or, in some versions, *puruṣa*), is identical for everyone. Thus the ultimate understanding of "self" is an understanding of "brahman," which is to say, an absolute that encompasses all apparently individual selves. In the case of classical Advaita Vedānta, this identity is spiritual. But in the case of absolute monism, it is equally a matter of being part of *prakṛti*, nature.

THE BIOLOGICAL AND THE SOCIAL

FLA: Perhaps we might delimit the territory by defining what we will talk about (and not) so as to provide a roadmap of sorts for our reader? I suggest we start with the most elemental and foundational structures or ingredients and then move to how these might inform our making and creating of narrative fiction.

At the most elemental and biologically material level, the neurochemical makeup of the brain and its complex web of physical electro-neuroactivity results in what we understand to be selfhood. Of course, this simple reduction (and very superficial overview of the work of Joseph LeDoux, among many others) is far from the last word on the matter. As with chemistry, so too here the substance we call the self grows from and is entirely different from the many ingredients that make up the self.[1] Roughly, cognitive neuroscience seeks to identify the neural processes that perform known psychological functions, and to explain these in terms of the work of specific neural systems, while eventually helping psychiatry to treat mental disorders. Jean-Pierre Changeux, Stanislas

---

1. See Aldama, *Why*, especially chapters 1 and 2.

Dehaene, and Lionel Naccache have proposed a limited but highly interesting theory of the self called the "neuronal workspace hypothesis." (See "The Global Neuronal Workspace Model of Conscious Access: From Neuronal Architectures to Clinical Applications.") It does not seek to solve the problem of consciousness as such, nor does it deal as yet with all aspects of consciousness, but as a research program it has a strong momentum. It posits that in the brain we can distinguish two main computational spaces: 1) a processing neural network "operating upon primary and motor stimuli, the contents of long-term memory (including a semantic database), the self and subjective personal experience, and systems of attention and evaluation involving motivation, rewards, and, in a general way, emotion" (Changeux, *The Physiology of Truth* 88); and 2) a global workspace "consisting of a distributed set of excitatory cortical neurons that are very richly interconnected" (88). This model is akin to what Mario Bunge calls in *Philosophy in Crisis* and elsewhere "emergent materialism," and, because it acknowledges that mental processes are strongly influenced by the social context, it can be supplemented by empirical developmental psychology and social psychology.

PCH: I largely agree about the biology of selfhood but with some qualifications. It is important to make some distinctions here. First, "reductionism" (the explanation of evidently "higher"-level phenomena, such as consciousness, through "lower"-level phenomena) is generally a good thing. But it is not clear that such a reduction is possible in the case of mind. Let me briefly recapitulate an argument I have made elsewhere.[2] It is quite possible to reduce any restricted set of mental phenomena to physical phenomena. For example, you could give a purely physical account of me. You could go on to give a purely physical account of LeDoux and his books and articles (his motor cortex produced such-and-such finger movements, producing such-and-such key presses, producing pages with such-and-such marks, etc.). But you could not continue the process all the way to yourself. If you tried to apply this to yourself, you would reduce your own sentences to bits of matter (ink) on other bits of matter (paper). Lacking intention (except as some sort of dorsolateral prefrontal activation pattern, conjoined with motor cortex activity, etc.), these bits of matter would lack meaning and thus truth conditions. Therefore, in order to make any claims about the mind, brain, or self, we need to leave something outside of physicalistic reduction—namely, the intentional subjectivity of the speaker.

2. See, for example, "Literature."

This argument is, of course, not without precedents. In effect, it is a version of the quantum mechanical idea that in order to describe any system, you need to place a "cut" between the observer and the system. Here, the system is the physiological reduction. The observer is what is presupposed by the reduction. In almost identical terms, it is what we find in Patañjali's Yoga philosophy. In Patañjali, the self is an "observer" of even the mind's own processes (see 2.17, 20, and 23). We also find a version of this in phenomenology in the idea of the transcendental ego. Each act of self-reflection makes the self into an object of consciousness. Once that occurs, however, the self-object is no longer identical with the self, for the self "transcends" self-objectification; it is, again, what observes the self-object.

Some people try to solve the problem by claiming that meaning is an "emergent property of the system." But this does no good. An emergent property is a well-defined idea in, say, systems theory. There it refers to a property that applies to a collection of elements—for example, symmetrical distribution across a space—but does not apply to individual elements (a single particle cannot be symmetrically distributed across space). In the context of mind "emerging" from brain, it is far from clear that the notion is well defined.

On the other hand, none of this says that intentionality is anything additional to the body. This "incompleteness of material reduction," as we might call it, applies globally, not locally. In any delimited system, the reduction is perfectly complete.

Having mentioned quantum mechanics, I should note, however, that my account here differs sharply from that of most treatments of mind/body issues in the framework of quantum mechanics. Most significantly, the idea of the epistemological cut is simply not taken all that seriously. Rather, it is assumed that the intentionality of the observer is causal with respect to the observed phenomena. (See Henry Stapp for a discussion of the standard views, by which "mind and matter . . . become dynamically linked in a way that is causally tied to the agent's free choice" [888].)

Finally, as to the material you have added on Changeux and others, I should mention two things. First, the processing network of Changeux cannot explain the self, since it presupposes the self. As you quote, the network operates on "the self and subjective personal experience" (Changeux, *Physiology* 88). Moreover, as just indicated, it seems to me a fundamental error to say that any neural process operates on "subjective personal experience" rather than on the neural correlates of that experience. I do not mean this as a criticism of Changeux. The technical distinction may not be important for his specific claims about neural architecture. But it is important for our understanding of the self. As to the developmental and other factors allowed by Bunge, these do not necessarily bear on biological reductionism. All such factors will have their neurological consequences through the usual biological routes. Thus, allowing developmental and other influences on the self—as everyone would—does not qualify biological reductionism, but rather rejects what might be called "strong genetic determinism."

FLA: As you and I both know well, there is the inseparable other side to this equation: the crucial component of the social. As we know firsthand and also explore in our work, the social in time and place matter significantly in the shaping of the self. This is what allows us to talk about how, say, in the time of the American Revolution such and such a personality emerged. I have the impression that a new personality or self emerged in the eighteenth century. Before this time, we did not have, say, a Benjamin Franklin, Thomas Paine, or Thomas Jefferson. These kinds of personalities are visibly new. And this perception that a new self emerges in new circumstances is not entirely objective or arbitrary.

PCH: I see what you are getting at here. But I would put it differently. First, I would isolate practical identities as individual sets of propensities, capacities, and ideas.[3]

One could reasonably identify a practical identity as the "self" of a person, and that practical identity would be instantiated in the brain. This is a different sense of "self" from the more minimal, momentary, philosophical sense I described before. Given practical identity at the individual level, we need to distinguish two things. First, there are some genuine impossibilities in prior periods—for example, no one had the procedural schemas for driving a car in the third century. However, my view is that these impossibilities are pretty banal.

The more interesting constraints are not on individual practical identity but on the systemic consequences of individual practical identity. Insofar as you are thinking of Franklin, Paine, and Jefferson as, say, egalitarian, I would not actually agree that this is a new personal identity. The same general ideas seem to arise continually throughout history. For example, there are egalitarian tendencies in many early Indic writings, ranging from the anticaste orientation of Buddhism to the revolutionary

---

3. For a more in-depth discussion of "practical identity" see Hogan, *Empire*.

zeal of the Sanskrit drama *Mṛcchakaṭikam*. The difference is that the social conditions were not such that the ideas and impulses were able to produce widespread, systemic consequences.

Here, I take a sort of Marxist view. On the one hand, there is broad social determination. But that social determination does not operate at the level of the individual. Sometimes Marxists act as if it does operate at the level of the individual. However, if that were true, it would seem to violate a fundamental principle of Marxist activism—"Class origin does not determine class stance." But this nondetermination does not apply to the entire system. If one thinks it applies to the entire system, one violates another fundamental principle of Marxism—the opposition to "voluntarism," the notion that social agents can simply join together and change the world, whatever the prevalent conditions of political economy. (On voluntarism, see Lukács, *History* 4, 124, 134, 191, 318, and 322. By the way, the extension of Marxism to systems theory has been developed by a number of writers. One forceful development of this sort is that of Immanuel Wallerstein.)

FLA: Are we thus biologically social and socially biological? This would make sense and allow us to understand more clearly why our existence in time and place is not static.

As biological entities we are always already born in a social environment, and all our senses and all our neurobiological equipment are trained and educated by our parents, caregivers, extended family members, and others who attend to our needs as children. It is our social environment, our social existence, our social being, that triggers and molds the functioning of all the faculties inscribed in our DNA, such as our capacity for language. In this sense, then, we are a social nature that is capable of knowing and changing the world by its social activity, and this social activity in its turn changes us as individuals and changes our culture and our society, partially and as a whole.

PCH: Absolutely. Among the questions that arise here are those relating to the degree of genetic determination. To a great extent, neuronal growth, for example, is "activity dependent" (i.e., it occurs in response to some input, for instance, some social experience). But the precise nature of that activity dependence is important. Is the growth pattern fully fixed? Is it strongly biased to one outcome (such that, roughly, something has to go wrong to produce another result)? Is it confined to a limited set of specific options (as in linguistic principles and parameters theory, where language learning is primarily a matter of setting predetermined parameters embedded in innate principles)? Is it simply a matter of having a range of possible degrees (as when musicians increase the density of neuronal connections in relevant areas through practice)? This makes a difference. As you know, I tend to view the brain as being much more plastic than, say, evolutionary psychologists do. But, at the same time, I am generally inclined to view the broad social outcomes as changing only in proportions, not in kinds. Consider, for example, child rearing. It seems likely that the same general types of attachment patterns occur in different societies and different historical periods. However, the proportion of each attachment type may differ across societies. (For example, the distributions of attachment types may differ in Japan and the United States, but the distributions would span the same attachment types.)

FLA: Our faculties of reasoning (deduction, induction, abduction), judging (distinguishing, separating, ordering, classifying), and evaluating (good/bad, right/wrong, tasty/unsavory, attractive/repulsive), as well as our emotions, our motivation (or will or willpower), and our intentionality (our capacity to plan, to have an action and its result in our mind before materializing it as an entity out there in the world)—these and very importantly our language faculty are all biological adaptations, are all biological products of evolution, are all biologically innate, genetically present in all healthy human beings.

As a part of the human genome, they are a feature of our biological (animal) nature. They are part of our nature, we being in our turn a part of nature, the social part of nature that has the unique capability to transform the whole of nature, to "socialize" nature, that is, to change nature in order to adapt it to our (human, social, universal) needs.

So, in that sense, these faculties are not only biological adaptations. They are at the same time social adaptations. Their *nature* is in fact social and biological.

I mentioned Ben Franklin among others; however, we can surmise that our more ancient ancestors not only "metabolized" the nature surrounding them but also transformed their own genome many times. Thus we can say that even our genome is both biological and social and biological, in the sense that we are biological beings through and through, but we are a biology that is capable of transforming all biologies (including our own) by means of our social activity.

PCH: You are raising important issues here. We may have a slight difference on the issue of adaptation. It is not at all clear to me that everything you mention is adaptive. Just to be clear, as you know, an adaptation is a genetic mutation that (in interaction with then-current bodily and nonbodily environments) produces some metabolic, behavioral, or other mechanism that leads to increased reproduction. For ease of analysis, we typically explain this increase by reference to a function that the mechanism approximates. For example, fear is clearly a product of adaptation. Specifically, we tend to flee from certain sorts of things. That helps to prevent us from being killed, and staying alive increases our chances of reproducing. We usually explain this functionally by saying that fear makes us flee dangers. However, the mechanisms here are distinct from this function—clearly, since we fear many innocuous things and do not fear many dangerous things.

There are three problems with the adaptationist emphasis of much current evolutionary psychology. The first is that it often passes over the distinction between mechanism and function. The second is related. It often passes over the existence of nonadaptive features. Finally, it frequently multiplies adaptationist explanations—explaining not only components but also complexes of components in adaptive terms. This relates to, for example, "literary Darwinism." If we can explain the likely function of both emotional response and simulation in evolutionary terms, then it is a violation of fundamental principles of simplicity in science to posit a *further* adaptive explanation for the conjunction of the two in literature.

You also touch on the issue of coevolution—the idea that our culture changes our biological makeup, which in turn changes our culture, all in adaptive ways. I am very skeptical about this. First, it seems that, for any given social trait, we only need the explanation in terms of culture without any reference to biology. For example, if something turns up across cultures, people tend to assume it is biological. But that is a non sequitur. There are many reasons something can recur cross-culturally. All that is required is that our biology be plastic enough to accommodate the cultural practice. For example, we do not need to evolve as cooks in order to cook things. We just need to have adequate neural plasticity to be able to learn to cook once the practice develops culturally (i.e., we do not need coevolution for cooking; we do not need cooking-enhancing genetic mutations). Second, even when something is not easily explained by culture, it is often hard to spell out in terms of coevolution. Indeed, discussions of coevolution sometimes appear to suggest that adaptive genetic mutations can be provoked by cultural conditions, which is, of course, bizarre (and something everyone would deny if faced with the idea explicitly).

FLA: What do we make of the fact that in spite of the commonsense evidence, there is a long tradition in philosophy, theology, and even science of serious scholars ascribing to selfhood a status that takes it totally outside the boundaries of scientific explanation, of suggesting that we will never arrive at a material explanation of the self? I think here of someone like Emmanuel Mounier (1905–50), who founded the magazine *Esprit* (1932) and articulated his Jesuit position in the guise of "a philosophy of engagement . . . inseparable from a philosophy of the absolute or of the transcendence of the human model" and whose "Omega point" identified the highest and most accomplished state of selfhood (135).

PCH: Well, there is a religious impulse here that begins from the premise that the self is an enduring soul. I do not believe that, but I do not find it absurd. In fact, I think it is something we can expect, given the human experience of self and human feelings of attachment. The primary motivation for such religious beliefs may not be a primitive attempt at explaining the world (as is commonly thought). Rather, it may derive from the emotion of grief and the nature of consciousness with its experience of loneliness.[4] In that sense, it is not something that one could readily oppose with science—and maybe one should not oppose it at all, since it may provide grieving people with comfort.

On the other hand, there is the rational version of the idea that I mentioned earlier. This does not make the self eternal (unfortunately). But it does remove one aspect of the self from scientific explanation—in principle, since that part is outside the system being described. (By the way, note that even in order to say this, I have to distort the point by making that observing self part of the described system.)

IDENTITY: CONTINUITY AND DISCONTINUITY

FLA: What do we make, too, of the seeming paradox of the discontinuous yet continuous self? Here I am thinking of the phenomena whereby I can say that I am forty-three years old and that I am Frederick Aldama—the

---

4. For more discussion on issues of grief and loneliness, see Hogan, "Literature."

same guy I cannot even remember but only have photos of when I was first born, then one, two, three, four, and so on years old.

There is a continuity from birth to death in our conception of self. Yet there is almost no continuity in the biological existence—the forty-three-year-old Frederick is biologically nearly completely different from Frederick at one or two or three years old. The whole organism is almost totally different. Yet I recognize myself continuously as being Frederick. I think too of the fact that my father before he began to suffer from diabetes, sleep apnea, and depression was a completely different biological person in his functioning. Yet, he still considers himself to be Luis Aldama. Likewise, my daughter Corina is not the same person today as a six-year-old that she was as a five-year-old. I perceive her to be the same and my care of her is the same, but she's not the same organism.

We are not the same—even at the molecular level—yet we conceive of ourselves to be the same person. I do not wake up in the morning, wash my face, and ask "Who is that guy?" in any ontologically significant way. We are changing constantly, yet we take the default position: to recognize our selves and others as the same. I am always certain that I am I. I am whom I am.... *Ego sum qui sum.*

It is a phenomenon that we observe worldwide: all children, worldwide, develop this idea of the "me," the "I," the idea of self-agency, of self-responsibility for one's actions. This is certainly genetically determined. (Only in cases of serious and rare illness or brain damage does this sense of self disappear.)

PCH: These too are important issues. Of course, in part we are simply dealing with a definitional question. We can use "identity" to mean "comprising all the same properties" (including constituents, time, place, etc.). Then I cannot step into the same river twice, as Herakleitos famously put it, or be the same person from moment to moment. We can also define identity in a way that allows for continuity—most obviously, in terms of certain sorts of causal relations.

A fundamental issue here is whether we are asking what a third person might count as the identity of an individual and what that individual himself or herself might count as definitive of his or her identity. As to the former, we can stipulate a range of criteria, many in keeping with ordinary usage. One obvious way of stipulating identity is by reference to bodily continuity (as when we say of a corpse that it is Jones). Another is bodily continuity and life (as when we say that Jones is no more, even though his body is there). A third is bodily continuity and consciousness or possible consciousness (as when we say that this body on life support is not really Jones). Another is continuity of memory (the sort of case you address). An analytic of this kind is valuable. But, in the end, our preference for one or another alternative is largely a matter of stipulation.

The more interesting issue is how someone understands himself or herself as having a particular identity. Here, it is probably necessary to distinguish different sorts of self-identification. As a first approximation, we might distinguish two. The first is the imagination and categorization of oneself as an object of thought. This is what most people seem to have in mind when they speak of self-identity. We might refer to this as "explicit" or "self-conscious" identity. Note that our sense of ourselves as self-objects is highly reductive. It tends to be a matter of categories—salient, functional, emotionally provocative categories (such as religion or ethnicity; the question "Who are you?" elicits the response "I am James Flaherty, an Irish Catholic," and so on).

The other type of self or self-experience is implicit. It is our sense of the uninterrupted process of thought or action. I would say that we tend to expand this implicit sense of self to the point where it encounters a contradiction. (This tendency to expand our sense of self is probably at the root of the Vedāntic identification of *ātman* with *brahman*.) Often that contradiction comes in the form of other people—as when I want to eat the last piece of cheesecake, but Jones gets to it first. Then I differentiate my self from Jones. But if Jones and I together rescue the last piece of cheesecake from Smith, I may say that "we" saved the cheesecake, thus suggesting a sense of shared implicit self. Note that this contrast can even take the form of a distinction between our sense of self and our own bodies—I (i.e., the self) wanted to run the final lap, but my body just would not let me; or I wanted to resist the cheesecake, but my hunger (now understood as nonself) got the better of me; or I tried to recall, but my memory (now nonself) failed me. The minimal implicit self is, roughly, working memory supplemented by current motivational states that enter into working memory.

FLA: How has our brain evolved to deal with this discontinuous yet continuous self as noncontradictory? What mechanisms are in place to help with this? I think of memory function. . . . And Michael Gazzaniga identifies an "interpreter" entity interpreting everything 24/7 that generates in each of us a sense of self (see his *Human*).

PCH: I doubt that we need to worry much about potential contradiction here. Everything that we identify as a particular is both unified and disunified. It is not just the self. The point holds for a tree (it is a tree, but it is also branches and a trunk—sometimes leaves or even flowers). There are, of course, empirical issues here bearing on the brain. But such identification is presumably part of the usual complex of brain operations.

FLA: To have consciousness of our selves, in this case of Frederick Aldama or Patrick Hogan, there must be brain mechanisms sending this information—sources telling the brain that Frederick is Frederick and Patrick is Patrick. Can we ask where this information comes from?

PCH: Yes, you are certainly right that there are brain mechanisms involved. But I suspect that the precise complexes of neural circuits change with the type of identity at issue. It is one thing to recognize one's face in a mirror and another thing to remember what one did yesterday. For example, the implicit sense of self is probably connected with lateral prefrontal operations of working memory and anterior cingulate cortex-related monitoring of task contradiction.[5]

By the way, it is important to stress that self-recognition and self-understanding are not a matter of Cartesian certainty, as we are often inclined to think. We make inferences about ourselves, just as we make inferences about others. (Though Descartes was probably right that we do not need to make inferences about our current working memory, to put it in contemporary terms.) It is, admittedly, pretty rare that we make a mistake about seeing ourselves in a mirror—though it is possible (and we can certainly make mistakes about photographs). But we get our memories mixed up with disturbing frequency (see Schacter 104–13), as well as our motives (see Nisbett and Ross).

FLA: To be able to identify a coherent self in our selves and others has other less heady implications. If our default position is a sense of a coherent and unified self, so too is this the case with us as authors and readers of narrative fiction. Characters are built on the basis of this understanding: that there is a self, a person, and that this self is unique.

PCH: Actually, I do not think our purely cognitive sense of uniqueness is very consequential. It is merely a sense of particularity. Of course, as a sense of particularity, it is consequential. (Imagine a mind that has only category-level thinking and no way of differentiating instances; it would

---

5. On working memory and lateral prefrontal cortex, as well as anterior cingulate cortex, see LeDoux, *Emotional* 276–77; on the anterior cingulate cortex and task contradiction, see Ito and colleagues 199; on working memory and consciousness, cf. LeDoux, *Synaptic* 190–95.

be pretty rudimentary.) But at this level, uniqueness is simply our sense that instances of a category may be differentiated by time, place, and variable features. In my view, the real force of "uniqueness" comes with our emotion systems. We feel that some particular is unique to the extent that it has a powerful effect on one or another emotion system, especially when that force goes against category-based responses. For example, we fear strangers but not someone who is familiar. Attachment is probably the system most bound up with a sense of uniqueness.

FLA: I consider myself unique, not just because I recognize myself when I look in the mirror but because of decisions I make and an awareness of a responsibility for my own actions and behavior. In this sense, I am the *author* of my actions. I make decisions and am responsible for (answerable to) my actions. My sense of self as unique seems to grow from this sense of agency.

PCH: I agree about one's sense of self being bound up with agency. Again, that is why nonself is defined by inhibition. (I am not sure I would call it "uniqueness.")

FLA: Authors and directors I tend to like make narrative fictions that turn upside-down my sense of codes that encourage behaving, or acting, responsibly; those like Christopher Nolan tickle me with characters—and stories—that shake a little the self-as-stable-and-coherent brain mechanisms. It is probably why I like the first half of *Crime and Punishment* (1866) so much. I do not agree in my own action to the killing of the pawnbroker, Alyona Ivanovna, but can relish this action in its fictional form.

SIMULATION AND THEORY OF MIND

PCH: My approach to fiction would be similar but perhaps not quite identical. Specifically, I see fiction as fundamentally the simulation of emotionally consequential goal pursuit.[6] The simulation involves us adopting the point of view of other people. But it does not necessarily involve us sharing a sense of self with them. (It may or may not; that is a complicated

---

6. For an extended treatment of literature and simulation, see Hogan, *How Authors' Minds Make Stories*.

issue.) This is fully continuous with simulation in ordinary life. I simulate what I might do tomorrow (e.g., ask my department head for some favor), and I simulate what other people will do (e.g., how my department head might respond). The evolutionary function of simulation is well known. It allows us to try out scenarios that may be beneficial or harmful, then pursue or avoid them. We pursue or avoid them based on "story" emotions, as we might call them—the emotions we feel with respect to the content of the simulation. For example, I imagine asking my department head for something (say, a different classroom), then I imagine him responding by requesting a return favor (say, me attending graduation this year), which leads me to imagine the consequences of doing as he asks (having to delay my summer travel). I feel aversion to this outcome, so I decide not to approach my department head about the initial favor.

In addition to this "story" emotion, however, there is some sort of pleasure in simulation itself. If there were not, then the evolutionary function would be entirely lost. We would simply avoid any simulation of aversive outcomes.[7]

Given all this, we would expect fictional storytelling to arise as a pleasurable activity.

FLA: I would consider simulation to be an expression of our causal, counterfactual, and probabilistic thinking—those mechanisms involved not only in generating hypothetical situations (scenarios played out in our minds with our bosses or stories generally) but also in the way we project a continuous sense of ourselves *in the future*. This self-identity (our "I" or "me") involves centrally what some have called our autobiographical memory; our capacity to recall events that happened in the past and to use that recollection to project our self toward the future.

PCH: Yes, this is, of course, crucial to simulation. Again, it holds not only for self-simulation but also simulation of others. Indeed, the operations of self- and other-simulation are fundamentally the same. We do not simply know how we are going to act; we come to understand through simulation, just as we do with others—and as inaccurately.

FLA: Knowing my state of mind and another's is crucial to our growing of a sense of self. Infants already ascribe intentions to others. They already

---

7. For discussion, see chapter 1 of Hogan, *What Literature*.

manifest the rudiments of what is called a "theory of mind." At the same time, the infant is developing the material sense of the world in which he or she exists.

PCH: This too is key. As you well know, theory of mind comes in two flavors—inference and, once again, simulation. Though initially these were seen as two competing accounts of theory of mind, most writers seem to agree now that both inference and simulation are aspects of theory of mind, though one or the other may be more important depending on the precise task (see Doherty 48). Emotion (e.g., understanding that Jones is sad) may involve greater simulation, whereas certain sorts of general knowledge (e.g., understanding that Jones does or does not know about debates on the theory of mind) may involve greater inference.

FLA: As we have talked about informally before, much has been made of the rather recent discovery of the mirror neuron system (MNS). To remind our readers, this is that set of neurons that become active in the brain's superior temporal sulcus and the Broca region (one language center) when we perform *and* observe action-oriented gestures that have intentionality and teleology in them. They fire (identify) the intention of the action or movement and its purpose or goal. To this extent, they are certainly an important part of our theory-of-mind capacity. Indeed, much has been inferred from this discovery about precisely how we empathize, interpret, and understand other people's emotional states, produce language, and seemingly now everything under the sun. What is your sense of this discovery?

PCH: This work is certainly of major importance, and there are many things one might say about it. Here are two. First, it is unsurprising that mirror neurons seem to bear on "everything under the sun" and even that some fairly wild claims are sometimes made about how they determine everything from language to ethics. However, it seems reasonable to assume that mirror neurons are important—at some level even necessary—to all forms of social learning. Our only access to the world is sensory. Before language, our access to other people's minds is extremely limited. Of course, we have inferential capacities. However, it is not clear that there would be any basis for our inferences if we did not have some sense of identification with the purposes of others. For example, it may seem that we could infer Jones is interested in the cheesecake because he is looking at it. But we can infer this only on the basis of knowing that people tend to look at things when they are interested in them. Thus the specific inference seems already to presuppose a general theory of mind. The obvious way of solving this problem is through simulation. But that just pushes back the problem—how do we get simulation with no prior sense of other minds? Mirror neurons provide, so to speak, the traction we need to get either simulation or inference started. In effect, they are a form of spontaneous simulation that allows for subsequent inference and effortful simulation.

The second thing to note is that mirror neurons seem particularly crucial in emotion. They are crucial in two ways. First, they do indeed seem to be the foundation of empathy, because empathy is impossible without simulation. In *What Literature Teaches Us About Emotion*, I distinguish between empathy and emotion contagion. The initial, spontaneous simulation of emotion is presumably closer to contagion. But it is probably a necessary condition for empathy proper. Second, mirror neurons are probably crucial for the development of enduring emotional responses as these emerge in childhood (probably a sort of critical period for emotion). Specifically, there is evidence that we acquire various emotional responses from caregivers (see, for example, Damasio, *Looking* 47). This acquisition is presumably based on mirroring responses.

EMOTION AND SELF

FLA: As you have already pointed out, emotion is central to our sense of self. Indeed, the recent research shows how we bind and process somatosensory information. It is the emotion center that triggers a reflex emotion in us in our everyday encounters with and response to the world: negative emotion to the threatening and unsafe and positive emotion to the comforting and safe. Either way, the stimulus is then processed by the reason or "executive brain" system. When we are watching a film, this reason system overrides the emotion center, telling the brain that we do not have to actually run from the living room or cinema.

This area of research on emotion is rich for those of us interested in the how and why of fiction. It seems to suggest that the way we process emotions can tell us much about why we may have like emotions when we experience fictional worlds but know all along that it is *not* reality. Both real-life and fiction-elicited emotion signals follow the same neurologic circuits from the brain's emotional system to its cognition system and then *diverge* in their effects when the cognition system determines what kind of response is warranted: to act or react when the information is identified as pertaining to real life and to stop or not initiate action when the information is identified as pertaining to the make-believe of the film, say. This is why when we watch a film, the emotions already triggered can be felt as intensely as the real emotions triggered in a similar real-life situation even though our executive center knows film is make-believe.

There is the all-pervading myth that somehow our cognitive and emotion functions are separable; that Spock is more rational because he does not feel. However, the constant stream of sensory stimulation and information (sound and image primarily) is first processed by our emotion system, or what biologists call the "limbic system," which includes the hypothalamus and amygdala.

I have been thinking about this in terms of video games—something that seems to be taking up much of the attention of all walks of life these days, and the topic of a new book I am working on. When marauding zombies try to get you in the video game *Resident Evil* (1996), rabid mutants gnash at you in *Left 4 Dead 2* (2009), or monsters crawl from spacecraft cavities in *Dead Space,* they activate our limbic system—our fight-or-flight response mechanism. We experience the reflex emotion of fear and the adrenaline kicks in. However, our brains are equipped with a conceptual filtering capacity; it is, *grosso modo,* a prefrontal cortex response that allows us to distinguish between the real and the unreal, or between fictional and nonfictional events. So when those monsters attack, the sensory stimulus is prefiltered as nonfiction by the emotion system, then resolutely distinguished as such by our reason system, or the executive cortex that includes the orbitofrontal, prefrontal, anterior cingulate, and motor cortex, for instance. The executive brain sets in stone, so to speak, this evaluation of the monster as unreal. All the while, another monster has jumped out at you that you must destroy, triggering once again the emotion system with its respective reflex emotion, its prefiltering of real versus unreal, and its signaling of the executive brain for confirmation. The emotion and executive brain centers trigger an appropriate body response—and the body reaction in turn can intensify the emotion experienced.

PCH: These are very important topics.

On the issue of why we do not run screaming from the theater when a lion appears on screen—I have held different views on this at different times. Part of what is happening is, of course, prefrontal inhibition, as you say. But I do not actually think prefrontal inhibition is that strong, nor do I think it has to be that strong in our response to a film. Part of what is happening is habituation (i.e., the reduction of emotional force through repeated exposure to an emotion elicitor), but that cannot be definitive, since full habituation would presumably mean that we would cease to have emotional responses to film at all. Critical-period experiences of safety during childhood film-watching may also play a role. In other words, having grown up with film, we have a set of critical-period experiences and associated emotional memories that qualify our emotional responses in the cinema (or the living room before the television). But that would not explain the response of adult viewers who did not grow up watching, say, horror films.

All these factors are probably relevant. However, right now, I am inclined to give particular weight to something else. Specifically, intensity of emotional response appears to be bound up not only with immediate emotion elicitors but also with very short-term projections into the near future.[8] Moreover, these projections crucially involve our sense of "peripersonal space," a very proximate space that organizes our feeling of bodily security and potential action (see Iacoboni 16). The various perceptual cues that arise in a movie theater tend to inhibit short-term anticipations that impinge on peripersonal space.

FLA: As seen in our lives and at all levels, the emotions play a central role; they are as essential for survival as cognition. Our cognitive and emotive systems are inseparable—even from the evolutionary point of view. You need this automatic knee-jerk reaction to survive, but you also need appraisal and reappraisal to survive. This develops organically; like language, it grows with you even before you are born.

PCH: As you suggest, the entire idea of opposing reason and emotion is, frankly, kind of silly—though one finds it all over the place. It is like opposing the circulatory system and the respiratory system. Emotions provide us with motivations. If we did not have emotions, we would never do anything. In the broad sense of "emotion," which includes hunger and thirst, the emotion systems are what lead us to function in the world at all. In contrast, processes of inference, categorization, and so forth (roughly, "reason") serve three functions. They help us to ascertain current conditions in the world, to infer future conditions, and to plot

---

8. See Hogan, "Sensorimotor Projection."

actions related to those conditions. Of course, this division is somewhat artificial since emotions enter at every point in our simulations. But these distinct systems and distinct functions remain.

This is not to say that I am an advocate of the view that emotions are great storehouses of reason. Emotions are evolved mechanisms. As such, they approximate functions in the environment of evolutionary adaptation. But they are not identical with those functions. We have various goals. Owing to the nature of our emotion systems, some emotional responses are very intense in dysfunctional ways relative to some of those goals. For example, I may feel intense hunger for cheesecake. At the same time, I may have a goal of keeping my girlish figure in preparation for the swimsuit competition of my local beauty pageant. Nonetheless, I scarf up the cheesecake. This is what people seem to have in mind when they speak about a conflict between emotion and reason. But this is not really accurate. Both my hunger for the cheesecake and my craving for the admiration of onlookers at the swimsuit competition involve emotion as well as inference, categorization, and other processes associated with reason. It is simply that one emotion is stronger—hunger, in this case (unsurprisingly, given that our emotion systems evolved at a time when simply getting enough calories was probably highly adaptive).

FLA: If I understand correctly, as I grow a balanced reason and emotion system in Corina, I'm growing a more sophisticated appraisal system that subordinates the impulse to be immediately satisfied by eating a bucket of Halloween candy in one sitting. This is to say, a balanced reason and emotion system does not achieve its full development, its full potential, outside of a social context. While infants arrive in the world with a charged emotion system—they cry when hungry or neglected, and smile when touched and fed—caregivers function as surrogate reason systems until the infant grows one of its own. They soothe and inhibit so that the little ones can *think* instead of reflex emote. As they grow, their emotion and reason systems come more into balance—they even begin to *think* about the emotions we experience. Working together, the emotion and cognitive systems allow us to ponder, assess, and modify our actions— and sometimes in ways that run counter to our reflex emotions.

Working together, the emotion and reason systems allow for the growing of our capacity to causally and counterfactually (and probabilistically) map our physical (objects and functions) and social (people and institutions) worlds. We get nice kickbacks here too. When we plan, put into play, and accomplish a goal, whether in relation to the physical or to the social environment, we are rewarded with neurochemical release of oxytocin and dopamine—the feel-good brain drugs.

PCH: It is certainly true that prefrontal emotion modulation matures. Moreover, such modulation does involve inference, categorization, spatial organization (e.g., recognizing where Mom is physically), imagination, and so forth—the processes that one might refer to as "reason." But I still hesitate to call such development a balance of emotion and reason. It is, rather, a growing ability to modulate response by reference to inference or related processes, *and to emotion*. In that sense, it is a balance between immediate emotional response and another, more complex sort of emotional response (which incorporates nonaffective cognitive processes as well). For example, children begin to develop mood-repair mechanisms that ameliorate aversive feelings (on mood repair, see Forgas 258). These mechanisms involve, for example, the recruitment of emotional memories to qualify the current emotional state, the imagination of pleasing outcomes, alternative and often more complex categorizations of both external and internal phenomena, and other processes. In other words, mood repair involves an integration of emotion and cognition throughout.

I should also mention that it is unlikely that emotion systems are fully developed at birth. Rather, it seems fairly clear that they develop through critical-period experiences, the aggregation of emotional memories throughout life, and perhaps other processes.

FLA: I think this is why I am so drawn to social neuroscience. Not only are we the most social animals among the mammals and primates, but we cannot develop biologically outside of the social, outside the existence of parents, family, teachers, and mentors. This is why emotions play such a central role in our growing of a self. Emotions are mechanisms involved centrally in the establishing (and cutting off) of relationships that ensure our protection in the social.

I mentioned the importance of memory to our sense of self in the past and our ability to project this self into the future. This imagining or hypothetical thinking is part of our growing of a capacity for causal, counterfactual, and probabilistic thinking. In our constant interaction with the natural (organic and inorganic) and social (people and institutions) world, we imagine and work through in our minds possible and probabilistic outcomes to actions and actually do the work to modify our environments and/or our expectations. As we modify our natural, personal, and social environs, we also get to know our own abilities better.

Because we grow in this capacity to formulate or perceive relations of causality, we therefore also automatically possess the capacity to perceive and formulate counterfactual hypotheses, arguments, and thoughts generally. We also grow and sharpen our capacity to create maps—of the human (social) and physical (natural) world—learning thereby to create new maps within the chain that allow us to consider new possibilities and formulate plans with probabilistic outcomes for what our situation will be in the world in the future.

PCH: I am sure you are right on the whole. But I suspect that, except in limited contexts, our capacities for inferring probabilities (or explaining causes) plateau pretty quickly. One problem is that there are just too many variables that we do not know about. It becomes very difficult to anticipate particulars in any detail. When one first begins to drive, one might be relatively clueless about what to expect from other drivers. It does probably take some years to acquire a good set of expectations. But I doubt that it keeps improving. I happen to be working right now on literature treating Kashmir. My sense in reading the scholarship about the history and politics of Kashmir is that there is agreement on a few basic points, but other than that no one has any idea what is going on and who is likely to do what when. Of course, that is a much more complex and unstable situation than the ones you have in mind. But I feel that most of life is only less dangerous in its unpredictability, not necessarily less unpredictable.

Moreover, it is not just complex, objective conditions that are a problem. Our capacities are limited. The point is obvious even with respect to the presumably easier task of explaining what has happened in the past. Our causal inferences have a basic functionality, of course. However, they remain simplistic and highly biased.

FLA: Many different experiments show that from a very early age children have everyday ideas about psychology, biology, and physics. Their brains are equipped from birth with the capacity to construct a surprisingly quite accurate picture of the way their surrounding world and its inhabitants work.

The awareness of self and body and the awareness of our interconnected lives follow the same neural circuit and make possible our physiological resonance with others and our capacity for empathy. The evidence from neurobiology confirms that the social informs directly the manner in which the brain develops its capacity to fully explore the world *as it is* (by coming to understand causality) and the way(s) it *might be* (by positing counterfactual scenarios); how the social from birth through adulthood shapes the basic *neurobiological* and *neurosociological* mechanisms required for further and deeper knowledge of the world as it *is* and as it *could be*. It is this "could be" that might offer us some insight into our fiction making and consuming activities.

STORIES

PCH: Certainly, our processes of simulation are guided by some sense of possibility. There seem to be a couple of alternatives here, bound up with the evolutionary functionality of imagination. We can simply imagine desirable outcomes. We may refer to this as "fantasy." But that has very limited evolutionary benefits—confined basically to temporary mood repair. Functional simulation requires that we sometimes imagine aversive outcomes. But simulation of aversive outcomes is not beneficial in itself. For example, it is not beneficial if it is random. Suppose I simulate going someplace to eat fruit. That simulation incorporates a memory of a predator lying in wait at the spot. My aversive "story emotion" will prevent me from going to eat the fruit. But that helps me to survive only if the simulation is more accurate than chance. If I imagine the predator wherever I imagine food, then it won't do me any good. So, clearly the imagination has to be constrained by some degree of accuracy. Moreover, that constraint seems to be connected with our pleasure in simulation. This is part of the endogenous production of reward for "correct predictions of the future" discussed by Peter Vuust and Morten Kringelbach (266). Clearly, the simulation itself is not immediately verified or falsified. However, constrained simulation (as opposed to fantasy) should be associated with emotional memories of correct prediction, which would have the same motivational function. That is the functional background for our predictive capacities in simulation. It also suggests why we often find a narrative emotionally ineffective if it includes ad hoc violations of causal principles.

There are, however, two qualifications to these points. First, the relevant mechanisms are undoubtedly limited in scope and applicability. If our ancestors were more accurate than chance in avoiding predators, it does not follow that our responses to the world are, in general, particularly accurate. Second, with respect to fiction, there is the strange fact that we can alter the principles guiding our simulations almost at will. I would be doing a very bad job of simulating my day tomorrow if I imagined that I could fly or magically transform my evil nemesis into a frog. But we do that in literature all the time. Moreover, this is not identical with "fantasy" in the technical sense. We can establish rules guiding imagination that we do not violate in the course of the literary simulation. For example, in Sanskrit literature, curses can be effective and they can be qualified or limited after the fact, but they cannot be entirely reversed. Put simply, the causal principles of a story need not be the causal principles of our ordinary simulations that are functional in the real world.

FLA: Building on your important point I would say that our causal mechanisms allow us to emplot stories but, because we exercise this along with our counterfactual and probabilistic reasoning, this in no way determines the final shape of the story. These innate cognitive mapping capacities grow in ways that express themselves in the making and consuming of narrative fiction in all its guises, including in video games, films, short stories, novels—and comic books.[9] Indeed, when we watch a film, we exercise the same cognitive mechanisms (causal, counterfactual, natural, and social mappings) already at work in young children who playfully invent storyworlds populated with imaginary companions and characters.

PCH: Absolutely. Of course, creativity in verbal art and elsewhere is connected with expertise. A fundamental principle of "creative cognition theory" is that creative work must be both novel and task appropriate, thus manifesting expertise. (For recent work exploring creativity, see the essays in Kaufman and Sternberg.) But creativity is also bound up with the more diffuse associations that characterize childhood imaginations.[10]

FLA: Just as the research on our growing of cognitive mapping mechanisms can shed light on our storymaking activities, I wonder if the research on personhood and the emergent functions that relate to this can enrich our understanding of how readers recognize a first-person versus a third-person narrator, or a flat or round character.[11]

PCH: Undoubtedly. This returns to the idea of predictability. Arguably, a round character has two characteristics that a flat character does not have. First, he or she provokes simulation. It is much more difficult to simulate the

---

9. See Aldama, *Your Brain on Latino Comics*.
10. On the issue of childhood associations, see chapter 3 of Hogan, *Cognitive Science*, and citations.
11. See Aldama, "Characters in Comic Books."

thoughts and actions of a flat character. Second, he or she resists inferential prediction. Conversely, it is much easier to use inferential theory of mind on a flat character. This is in keeping with the sense of subjectivity that goes along with character "roundness." A round character is like we are for ourselves. As is well known, we see ourselves as responding flexibly to circumstances, whereas we view others as acting from fixed character traits (see Holland, Holyoak, Nisbett, and Thagard 222–24). The genius of authors creating a round character is that they manage to imagine a character as "selflike" in responding flexibly to circumstances.

FLA: In my work I put certain limits to how I consider theory of mind operating within the narrative fictional space; it is and always will be an artificial construct. Jane Austen's characters that misread other minds to elaborate degrees will do so for now to eternity. We as readers simply follow the author's scripts and cues. Whereas in everyday life, our use of theory of mind is a rather messy business; we often misread interior states of mind in our everyday activities. In narrative fiction there is no guesswork.

PCH: You seem to be suggesting some disagreements with Lisa Zunshine here. As you know, I disagree with Zunshine about the degree of theory-of-mind embedding to be found in literary works (character A believing that character B imagines that character C anticipates that character D, and so on). I also feel that she leaves aside many factors when she treats "why we read fiction." But I do not agree with the criticism of her work on the grounds that characters are not real people. That makes some difference, of course. But the difference is, in my view, very limited.

I take it that you say "in narrative fiction there is no guesswork" because there is no fact there—or, rather, there is no fact unless the author explicitly tells us. But I do not believe this follows. There are at least two sorts of fact about unstated properties of characters' minds, and both bear on our usual simulative (or inferential) theory-of-mind processes. First, there are general facts about the world. We may reasonably say that if someone in the real world acted in this way and said these things, then we could conclude that he or she had such-and-such psychological properties or made such-and-such inferences; therefore, it is reasonable to impute the same properties or inferences to the character in a novel who acts in this way and says these things. The point applies to embedded theory of mind, thus one character's understanding of another character. For example, I recently watched Ralph Fiennes's *Coriolanus*. That production suggests particularly strongly that Coriolanus has a very deep, but also troubled ("insecure") attachment bond with his mother. Moreover, when he agrees to pursue a peace treaty with Rome, it appears that he is profoundly concerned about just what his mother's feelings are about him and how they will be affected by his decision. This is a theory-of-mind concern. Clearly, Shakespeare did not state any of this explicitly. But it would be strange to deny the validity of this simulation—that is, the viewer's simulation of Coriolanus's simulation of his mother's feelings and intentions—or an inference that supports such a simulation. Conversely, it would be strange to deny that this has the same type of uncertainty or ambiguity as our usual "messy" theory-of-mind simulations.

The second sort of fact bearing on characters concerns how the author himself or herself simulated the character. Note that this does not involve precise labeling or even self-conscious judgment. Neither Shakespeare nor Fiennes needed to know anything about attachment theory for one or both of them to have simulated Coriolanus as having profound but insecure attachment to Volumnia and a response of something like panic to the thought that she would withdraw her love (and perhaps die in the course of continuing war).

In short, the theory-of-mind situation in literature seems to be largely the same as it is in life, with two limited differences. The differences are that we sometimes know more about characters than we do about people and that we sometimes have to simulate the author's simulation of the character, not merely the character alone. We are particularly likely to shift from general facts to author simulations when the author was writing in a cultural or historical context different from our own, thus rendering our life-based generalizations or associations less trustworthy.

FLA: Perhaps I was not clear in what I meant by there being no guesswork. To put it very briefly, an author in possession of emotions, intelligence, personal and social experience, and so forth, does her educated guesswork and uses many other mental operations to formulate theory-of-mind hypotheses in real life. On the basis of this experience, he or she knows how and when and in what guise to ascribe to his or her characters a theory of mind. Like in all other instances of creating narrative fiction, the author uses aspects of reality as building blocks for the invention of storyworlds and characters in possession of all sorts of psychological or mental traits and forms of behavior. That said, a reference to theory of mind may offer a useful analytical tool for us to understand how an author creates characters and their behaviors (in a specified way) as well as how readers use theory of mind to engage with a work of fiction . . .

PCH: To some extent, then, it seems that our disagreements are superficial and terminological.

FLA: More generally, the scholarship on fictional characters today one way or another also appears to conflate the fictional with the real, the constructed character with the flesh-and-blood human being. I think here of Alex Woloch's *The One vs. The Many* that seeks to give nuance to E. M. Forster's flat and round character taxonomy in his formulation of a typology of character (or character system) that is based on the character's minor or maximum presence in storyworld spaces. Enough is said with this, however, so I'll move on.

While we share universal biological mechanisms that together allow for the growing of the self, our infinitely rich and complex social environments shape the expression of this biology in radically different ways. It is why we cannot predict or establish causal laws that describe how millions of people will react in the face of any aesthetic experience.

Our radically different selves grow from a shared biology. It is a fact established now by biolinguistics and neuroscience that there is what Chomsky calls a faculty of language—a brain-based faculty universally present in human beings. We all have this faculty, but not everyone uses it the same way and with the same skill. Goethe had it, but so did Hitler. Goethe wrote the *West-Eastern Divan* (1814-1819) and *Faust* (1808), and Hitler wrote *Mein Kampf* (1925). Goethe favored the arts and science of his day, and Hitler favored capitalistic barbarism. Faulkner wrote *Absalom, Absalom!* (1936) and George W. Bush could hardly get a sentence right.

POLITICAL AND ETHICAL IMPLICATIONS

PCH: Responses to fiction clearly differ to some extent. In part, this is a matter of personal experiences, prominently emotional memories.[12] There are also differences in encoding sensitivities (when my students read Indian works there are culturally key moments that they do not even notice).

---

12. See chapter 2 of Hogan, *The Mind and Its Stories*.

Inclinations to engage in effortful simulation also vary, as do inhibitions of empathic response due to identity categories. These and other factors work against uniformity of aesthetic response—though some of them are alterable. For example, one can enhance one's encoding sensitivities, decide to engage in effortful simulation, self-consciously modulate in-group/out-group prejudices, and so forth.

On the other hand, we may disagree about radical differences in social environments. It depends what you have in mind. If you mean different cultures, then I would demur. Again, cultures may differ in the precise proportions of various attitudes, ideas, propensities, and other matters. But they do not seem to differ on which attitudes, ideas, propensities, and so forth, are present. The remarkable recurrence of the same literary genres (e.g., romantic tragicomedy) in culture after culture suggests this. Of course, if you mean that particular, individual conditions can vary radically, that is true. Although even here I would not say that this is the usual case. There are certainly variables in how people develop emotionally, for example. But I suspect that the variability has a normal, bell-curve distribution, with radical differences only at the tails. These would largely be pathological cases.

As you know, I have written about both Hitler and Bush.[13] But I cannot claim to understand them, especially Hitler. In a sense, *Mein Kampf* falls within the normal distribution since it is a fairly standard sacrificial emplotment of history. It combines this with a particular form of categorial identification and in-group/out-group division, one involving a high proportion of disgust motivation. The particular historical circumstances in which he was born not only contributed to this combination but also contributed to his success. In other words, it was a matter of the system in which he found himself. With even slight changes in that system, Hitler would simply have been a crank. In these ways, then, we have a common phenomenon in the operation of complex systems. A particular set of small variations at the outset (the nature of his sacrificial emplotment, the precise emotion systems involved in his in-group/out-group divisions, and so on) produce catastrophically different outcomes.

FLA: So do we throw in the towel and say that it is all a matter of opinion based on personal experience? We cannot know how each individual will actually experience the novel, film, and so forth, but we can generate what

13. See *Understanding Nationalism*.

the text's ideal audience response would be by sleuthing out the respective devices and structures that make up the narrative blueprint. This is where I think the approach and aims of narratology can be useful.[14] It can provide concepts and instruments through which we can study the blueprints created by authors, artists, and directors that circumscribe the phenomena of narrative fiction making and consuming.

PCH: I agree. It even helps to some extent with cases such as Hitler. *Mein Kampf* crucially involves emplotment. Examining that emplotment gives us a better sense of the work's effects. Similar points could be made about its target audience, though presumably our aims in this case would be rather different from isolating, say, Faulkner's ideal reader.

FLA: Narratology and the advances in the cognitive and neurosciences of memory, empathy, theory of mind, causal and counterfactual social and physical mappings, and much more can offer a foundational analysis of how an author or artist is able to create effective blueprints—effective in the sense of being able to have an impact on an audience.

PCH: I of course agree—though, with my usually pessimistic attitude, I would add that it is the effectiveness not only of artists but also of demagogues.

FLA: The self, agency, action, and responsibility are central to what has been called ethics. Within the tradition of the study of human behavior and action we have three main threads: virtue ethics (formation of a character aspiring to the good), deontology (duties or rules), and consequentialism (consequences of actions).

I am interested in what we might identify as a fourth strand, one that shares aspects of the virtue and consequentialist ethics. It is an ethics that says that all human beings are part of nature; they are part of the animal kingdom. They are special, social kinds of animals but animals notwithstanding, and therefore they all share the same genetic neurobiological architecture. It understands that humans in their activity (always a social activity) develop norms of behavior (conduct) that are also social; these could refer to anything from table manners, hygiene, and grooming to not committing adultery or murder. Among these norms also develop norms of interaction of one toward others: specific kinds of social norms we call moral norms—or simply morality.

14. See Aldama, *A User's Guide to Postcolonial and Latino Borderland Fiction.*

PCH: It is interesting that you pick virtue and consequentialism from this triad. It relates to our partial disagreements about prediction. In general, I am Kantian in ethical attitudes. In other words, I would favor basing moral decisions on evaluations of the intrinsic morality of the action (in light of moral principles—such as *always treat others as ends in themselves and never as mere means*). I generally accept the criticism that consequentialist ethics relies on impossible predictions. On the other hand, I do qualify Kant with respect to some short-term consequences that are predictable with a high degree of confidence. In other words, I see the value of consequentialism as severely limited by our poor capacities at prediction.

Virtue ethics tends to enhance our tendency to think of people as good or bad. This seems to me to have unfortunate consequences. I would much prefer to confine claims of good and evil to acts.

FLA: I have kept at bay the deontological position, Patrick, simply because it appears to me to allow too much room a *faith-in* position concerning social justice and the like. I think either way we look at it, we can both agree that our existence as social, productive beings requires work and action to modify our own self (nature) *and* our societies and relations within society. This transformative capacity in time and place also modifies our codes of behavior *in time and place.* Human behavior and its codification that appears in *The Iliad* (ca. 1194–1184 BC) or *Gilgamesh* (ca. 2500 BC) is not the same codified behavior today.

At the same time that there is a great variety of sources for ethics, we have the whole domain circumscribed by frontiers established by our socio-neurobiological nature. This is what allows us to understand why we are essentially, fundamentally, a single species and why morality or ethics, notwithstanding its myriad particular manifestations, is at the same time a universal phenomenon. The basic rules of ethics are essentially based on what we are neurobiologically and the way we maintain our existence, our lives, through *social life.*

PCH: I wish I could say that the "heroic" ethics of *The Iliad* are absent today. It seems, rather, that they are all too prominent. Happily, so are the ethics of Euripides' *Trojan Women.*

FLA: As an infant learns to map its social and natural world, it is also fulfilling some basic needs in order simply to exist: need of affection and security and protection, of food and shelter and clothes, of learning, of beauty and knowledge and know-how, of social collaboration, altruism and fairness, of happiness and laughter, love and touch. There is also the sense that the 6.6 billion people on Earth today are almost identical from the point of view of the early cognitive development, anatomy, and the physiology and the functioning of their brains.

PCH: Yes. I would only add that many things enter here—contingencies of child rearing, continuities of the physical environment, patterns in group dynamics and complex systems. Due to these and other factors, the conditions into which children are born are largely the same as well. More precisely, these and other *nonbiological* factors constrain the range of variability in those environments.

FLA: Our most basic neurobiological mechanisms such as our capacity for empathy and mind reading are involved in our codes of behaving and our moral judgments. It seems this is necessarily involved in our saying that such and such an activity is wrong, harmful, or should be punished. In a court of law today, we as a jury are asked to determine the intent—the interior state of mind—of the defendant. We need to determine that person's state of mind and in so doing also ascribe *responsibility* or not; hence, too, certain protections that are in place to protect children, whose brains are at a different stage of brain development than those of adults: adults have rights together with responsibilities, whereas children have protections, rules that protect them.

We are the only social part of nature that is *capable of transforming through work, through our social being, the totality of nature,* or at least that part of nature contained in the planet Earth. And by doing so, we are exclusively capable of bringing in factors that continue our evolution as part of nature in ways that change our selves. We have needs just as beavers and ants have needs that they must satisfy in order to survive, but no other animal society does what humans do: by satisfying our needs, we *transform those needs* and even multiply them.

PCH: I know what you are getting at here, but I am not sure that we actually transform our needs. This goes back to the idea of coevolution, of which, again, I am skeptical. I am not saying that we are simply born with all our needs. A child bonds with a particular person after birth. Then it is entirely reasonable to say that he or she needs that person. In other words, the child may suffer physical decline if separated from that person.

Similarly, various addictions alter our needs (indeed, romantic love and attachment appear to be connected with addiction[15]).

Of course, we do change the means by which we achieve ends. If one wants, one can refer to those means as "needs" also. If we need food and the only means to get food is money, then one could say that we "need" money. Nonetheless, I feel it would be clearer to say that the need itself hasn't changed but rather the means of achieving the needed outcome.

FLA: What I find so exciting and scientifically powerful about the findings in cognitive science and neurobiology is that they have established as a solid fact that we as humans are one; that historically (prehistorically) we all share a common origin; that we all came from that initially small group of walkers who went from Africa to the other continents spreading more and more of their common genes, their common genome.

And that this fact of our common humanity is the strongest support for the view that all of us should have the same rights and opportunities to develop fully our cognitive and emotive selves and to realize our causal and counterfactual potentials in the form of science, narrative fiction, or anything yet to be imagined.

PCH: There is nothing I can add to your eloquent statement, with which I wholeheartedly agree.

---

15. See Hogan, *What Literature* 83–83 and citations; Stein and Vythilingum 240; and Panksepp 54.

CHAPTER 2

Verbal Art
and Language Science

PATRICK COLM HOGAN: It goes without saying that literary study has been deeply indebted to the study of language, particularly linguistics, since at least the first wave of structuralism. This is unsurprising, in some ways quite intuitive. After all, literature and orature are made out of language. Given this, it seems that no discipline could provide more appropriate and effective theoretical tools for literary study.

On the other hand, the rise of linguistic approaches to literature has made the limits of such approaches painfully obvious. This is true in several respects. First, whenever there is some sort of "master discipline," other potentially important areas of study are likely to be ignored. For example, literature involves language, but it also involves emotions, imagination, and other psychological processes. Mainstream narratology seems to have been so dominated by linguistic ideas that it has largely ignored affective science or theories of simulation—even as it has expanded to treat non- or only partially verbal media, such as film and graphic fiction.

FREDERICK LUIS ALDAMA: Yes; although I am working to establish a unified theory of aesthetics, I think that the idea that one discipline (in this case linguistics) can account for everything related to the domain of literature is untenable. Of course, there has been a strong impulse to turn to linguistics precisely because literature is expressed in words through language. This said, while we clearly access characters and their worlds through the medium of language, we need also to attend to how the emotions and ethical moral attitudes are built into the characters we consume as well as other important psychological processes involved in the making of narrative fiction.

I mentioned already in chapter 1 our causal, counterfactual, and probabilistic faculties—areas of study that are generally bypassed for one reason or another and ignored in the study of fiction. (Alison Gopnik studies these faculties in her coauthored 1999 book, *Scientist in the Crib*, and in the single-authored *Philosophical Baby*, where she forcefully shows how children grow theories of the social and natural environments.) Such causal and counterfactual mechanisms are at work when the author chooses to invent a character with such-and-such emotion and ethics system. Whatever we find in the character's mind the real-life author placed there.

Of course, no author, no human being to be more exact, functions only in terms of emotions or only in terms of reasoning or in terms of moral attitudes. We function as a whole. To think in a way that puts aside the concept and experience of causality is only possible through the use of another mental faculty: our capacity to conceive counterfactual situations and concepts.

Our mind works as a whole, as you well know. If I have a jumble of ideas and images in my mind as an author, then it will be a jumble of ideas and images that the reader will find in my stories and also in the minds of the characters. I am the author. I put whatever is there *there*. I can very consciously put into the minds of my characters beliefs, thoughts, emotions, and attitudes that I do not share as an author. But I put them there.

It is inevitable that narrative fiction will be built with the building blocks that come from every aspect of reality that has entered the author's mind and that the author will put into his stories and characters. It is unavoidable.

This is why this myth associated with Hemingway—created essentially by media such as *Life* magazine, sensational biopics, and the like—that continues to spin its web in some creative writing workshops asserts in one way or another that you can only write well that which you know from firsthand experience.

This is nonsense—at the level of both the creating and the co-creating. Imagine a teenager living in a small town in Mexico surrounded by tropical forests who reads Dostoevsky; she reads about the terrible cold the protagonist feels in St. Petersburg, about the snow, the iced-over rivers, and so on. She has never had any direct experience of snow, yet she is able to imagine fully the scenes, successfully filling in the gaps in the blueprint. This concerns the visible part of our natural reality. However, she might also read a story written by Tolstoy that concerns financial transactions, marriage, and a guy who is dying that build all sorts of emotions into its blueprint, including, most importantly, those that surround the act of dying. Yet, this teenager is only at the beginning of her journey through life. If the Hemingway myth (in whatever iteration) were true, such a reader would not be able to experience the emotion system of the story. Of course, this concerns the act of reading. We can also say the same of her choice to write a story. This same teenager might choose to write a story that has nothing to do with her proximate experience—a story set in a Nordic climate about a man on his deathbed, say—and she would have the cognitive and emotive faculties to do so.

This is a hypothetical, but we can see such a refutation of the "you can only write what you know" myth many times over with examples from world fiction authors. Think of Carlos Fuentes in his mid-twenties, who did not have any direct experience of the 1910 Mexican Revolution yet was able to write one of the most magnificent novels of the Revolution, *The Death of Artemio Cruz* (1962).

This is to say, firsthand experience *comprises* not only that which I have touched (or that has touched me) physically directly but everything that I have read, discussed, seen in motion and in static pictures, and so on. It involves all my aesthetic capacities and feelings; all the education of my senses through direct experience (through direct acquaintance) and indirect experience (indirect acquaintance).

PAST USES OF LINGUISTICS IN LITERARY STUDY

PCH: Another difficulty with linguistic approaches to literary theory is that they tend to rely on historically particular linguistic theories, often theories that are no longer current in linguistics. This is not merely incidental. It tends to result from ordinary disciplinary developments. Theoretical work in linguistics is often highly technical, and it tends to become more technical as it develops within any particular theoretical tradition. Researchers outside of linguistics are often ill prepared to understand technical developments. In consequence, they are more likely to be attracted to theories that are available in more accessible summaries (usually older theories), as well as the early stages of theories.

The most notorious case of this is Saussurean linguistics, which entered into literary study in the 1960s, long after it was understood by most linguists as inadequate to explain or even describe fundamental linguistic phenomena. In *The Politics of Interpretation*, I discuss how many of Derrida's conclusions about meaning—its differential and deferred character, its relation to binary oppositions, and so on—are not a working out of instabilities in language. Rather, they are a working out of some inadequacies in the Saussurean account of language—specifically, an account of phonology generalized to semantics. In this way, one can see almost fifty years of mainstream literary theory as suffering from an unfortunate assumption of implausible linguistic principles.

FLA: I will admit that in my younger days as an undergraduate at UC Berkeley I found extremely attractive the systematic and deep-level approach offered by scholars who used a linguistic approach to the study of literature, such as Ann Banfield and Julian Boyd. The more I worked in this area, however, the more I realized its shortcomings.

I wonder if it might be beneficial to consider viewing this from yet another angle. To give us a chronological signpost, after World War II we see a rich variety of linguistic theories appear. Most of these post–World War II theories were developed in France, England, and the United States. Before this surge of diversity, there were essentially a handful of linguistic theories—all of them with a strong family resemblance and all of them one way or another related to Saussurean linguistics, or, more accurately, the proposals made by Saussure under the domain of linguistics. There was no big clash or contradiction between these theories.

Before we talk about the different theories, however, I have to say a few more words about Saussurean linguistics. As a young man in 1878, Saussure published his dissertation on the primitive vowel system in Indo-European languages, which earned him a lot of prestige. This thick book (four hundred pages or so) is about vowel systems in what was called at the time the Indo-European languages. It is perfectly coherent and consonant and inscribes itself easily within the dominant school of linguistics in the whole of Europe formed by the New Grammarians, essentially concentrated in Germany, France, and Britain but that also sprouted up in the United States with the German expatriate Franz Boaz, the teacher of Bloomfield and Sapir.

There were, of course, important scholars before these fellows who made important progress in the study of linguistics, studying language in a quite systematic way and establishing regularities in Sanskrit, for instance; this in contrast to others who mixed myth with equal doses of speculation in the quest, for instance, to discover an Adamic language. However, it was the New Grammarians that contributed something very important: a scientific approach to prove the global hypothesis of the existence of a protean mother lode for all Indo-European languages. They were the first international school of linguists to have a scientific methodology and outlook that set aside all attempts at explaining linguistic phenomena by theological or politically motivated pseudo means.

One of the most prestigious Francophone New Grammarians was the very young Saussure. The publication of his dissertation when he was twenty-one years old suggested great promise for the development of the field of linguistics internationally. However, he knew that he would have to redirect his New Grammarian training and scientific (empirical findings and results accompanied by observation and the discovery of regularities) impulse elsewhere if he were to take the study of language to the level of generality of a science of linguistics. He asked himself what language was and how it functions, operates.

The product of the work that grew from this can be found in the notes (turned into the *Cours de linguistique générale* [1916]) from his courses on "general linguistics" taught at the University of Geneva. Here he discusses the foundational problems I have already mentioned: what language is and what the proper domain of the study of linguistics is.

Before Saussure's notes were published as a book (the diligent work of his students), his theories had already caught the eye of Roman Jakobson and Nikolai Sergeyevich Trubetzkoy—who later wrote an extraordinary book on philology. It was Jakobson who met with members of the Moscow and St. Petersburg circles of the Russian formalists and explained what Saussure had done. Through Jakobson and the work that Trubetzkoy was doing in phonology, scholars interested in the scientific study of literature were given a model for the scientific foundation for the study of literary phenomena.

The first attempts by the formalists failed because they tried to use the *parole/langue* system as a way to identify the differences that make a difference between the language essentially of poetry (its literariness) and that of everyday usage. The attempt to directly apply Saussurean linguistics to the study of literary texts (essentially poetry) quickly led to dead ends, although the formalists also increasingly studied prose narrative fiction. Indeed, in their turn to prose narrative fiction, they dropped the strictly linguistic approach and followed instead the story versus discourse approach. Of course, the Stalinist bureaucracy cut short the life of this research program; it wasn't until Gérard Genette's student Tzvetan Todorov translated the formalists into French that their work on this score was rebooted.

The ambition to make this linguistics a more solid, scientific discipline—even formalized in the sense of formal systems such as logic and mathematics—took root and grew many branches in other places such as Czechoslovakia and the Nordic countries, particularly Denmark. This ambition to make linguistic structuralism something much more scientific and much more susceptible to verification and refutation took the name "glossomatics." (See the work of Louis Hjelmslev and Hans Jorgen Uldall.) In the United States, building on the work of Boaz, Bloomfield and Sapir began to establish regularities not of one mother-lode language of European languages, but of two big mother lodes. This is why American structuralism in linguistics shares many traits with European structuralism and at the same time has many of its own unique features.

This leads me to Zellig Harris, the post–World War II scholar who made important attempts at formalizing (mathematizing) linguistics and who was the direct teacher of a very young Noam Chomsky. We see in Chomsky's work the use of tools developed by Harris in his first stab at what he called the transformational generative grammar; transformational grammar was a tool developed by Harris in the analysis of syntax. In his very first version of his theory, which he later returns to and refines in the 1980s, Chomsky uses this tool.

Patrick, you mention that linguistics has become a very specialized field. You are right. In fact, today it is very difficult to follow and read. This impulse was accelerated greatly with Chomsky. As a disciple of Harris who had introduced a mathematical methodology in linguistics, particularly in his use of transformations, Chomsky made linguistics as a whole a more and more formal discipline—a more and more mathematized discipline. Already in this early work, we have to know quite a bit about mathematics to understand the theory.

Another part of the increased complexity of the discipline is that Chomskyan linguistics has evolved in several directions. Chomsky's original proposal made in the 1950s is very different from versions of it that you find today, for instance. The discipline as a whole has never remained stable or had the same aspect over the years. Scholars in English departments may think that they are applying a cutting-edge theory of linguistics, when in fact what they are applying is a very outdated linguistic theory; even Chomsky's own work has evolved radically from his formulations in the 1950s.

All this aside, there remains the issue of applying linguistic science as such to the study of phenomena that are not linguistic; that do not pertain to the domain of the science of linguistics.

We all use language in our everyday communication and find language in many places—books, newspaper, movies, television, you name it. Literature is made with language. But the proper study of linguistics is not language. Its proper study is not language use or linguistic use—it is the study of the faculty of language. Otherwise stated, what linguistics has to explain is not *how* Spanish works and much less how Frederick Aldama's Spanish works; the domain of linguistics is not the study of any particular language. The domain of linguistics is all languages in the most abstract sense of the term. Therefore, the domain of linguistics is that which is *essential* to the language faculty. (The proper study of the theory of gravity is the minimum essential traits of the phenomena of gravitation that we find in any and all forms of matter.)

LITERATURE AND THE SCOPE OF LINGUISTIC THEORY

PCH: There is another potential problem with the use of linguistics in literary theory that relates to the level at which the application occurs. The most basic sort of application occurs when linguistic and literary study overlap. This overlap is limited, but it is not insignificant. It tends to occur at, so to speak, two ends of linguistic inquiry. Specifically, it is common to divide linguistics into a series of subfields. At the most minute level, we have the study of speech sounds in phonetics and phonology. Research in these areas—such as the work of Paul Kiparsky, Nigel Fabb and Morris Halle, and Geoffrey Russom—has guided some enormously valuable work on universality and variation in principles of meter or other aspects of verse.

FLA: There is also the work of Carlos Piera and Bruce Hayes.

PCH: At intermediate levels, we find morphology and lexical semantics, treating meaningful units (largely words), then syntax and sentence-level semantics. Finally, the study of language above the sentence level is discourse analysis. Discourse analysis encompasses such practices as storytelling. Linguists tend to be more concerned with conversational stories than with narratives of verbal art. Nonetheless, some very important work in narrative theory has been done directly in discourse analysis and related areas, such as sociolinguistics. One towering figure in this field is William Labov. His model of conversational storytelling has been a major contribution to our understanding of narrative generally (see Labov and Waletzky). There have also been some subtle linguistic analyses of more clearly linguistic features of discourse, such as free indirect discourse, as in research by Sharvit.

FLA: Or Ann Banfield.

PCH: As these examples suggest, the majority of consequential work in direct application has been by linguists. This is unsurprising, at least in the case of sound analysis, since there have been great technical advances in phonetic and phonological analysis. It is difficult for literary critics without training in formal linguistics to apply this work. The major exception to this comes from one narrow aspect of semantics—the study of metaphor. On the other hand, even there it seems that, from the literature side, real theoretical contributions have been made largely by Mark Turner, who has a breadth of scientific training rare among humanists.

FLA: This reminds me, Patrick, of a time in France when a group of linguists, philosophers, mathematicians, biologists, poets, and fiction writers affiliated with the cultural journal *Change* (1968–85) gathered together their disciplinary forces in an attempt to apply linguistic approaches and methods to the study of poetry and fiction; they were particularly interested in applying Chomsky's generative linguistics. Among its members were Mitsou Ronat, Jean Paris, Leon Robel, and Jacques Roubaud. (One of its members, Jean-Pierre Faye, worked to push against the most ideologically charged forms of structuralism represented mainly in the later Barthes and in the Tel Quel group and therefore against the novelist Philippe Sollers and Jacques Derrida.) Ultimately, the product of the labors of the members of "Change" was rather small. Arguably their biggest impact was the publication of what became a best-selling book-length interview between the linguist Mitsou Ronat and Chomsky, *Language and Responsibility* (1979; reprinted in 1998 in *On Language*). In this interview, Chomsky tellingly remarks how "ever since the ancient Greeks people have been trying to find general principles on which to base literary criticism, but while I'm far from an authority in the field,

I'm under the impression that no one has yet succeeded in establishing such principles . . . That is not a criticism. It is a characterization, which seems to me to be correct" (*On Language* 56–57). In this dialogue and elsewhere, Chomsky identifies clearly the benchmark: to find general principles in the study of literature, to obtain results that are empirically verifiable and that possess a general explanatory power.

One of the big achievements of Chomsky is that he introduced for the first time the notion and imperative of *making absolutely explicit* the goals of the science of language. Very early in his work, he gave a mathematical and logical formulation to the goals of this science. He left no ambiguity in his formulations, no room to squeeze in pseudoscientific goals by those inclined toward obscurantism.

After Chomsky's clarification, most philosophy of language became irrelevant, and so too did much applied linguistics. Scholars were forced to clarify and be explicit about their goals or else face expulsion from the field of philosophy of language. Derrida was an exception; he was never explicit about his goals in anything; and one has to be very attentive to his interviews and writings to discern his prelapsarian linguistic impulse.

I agree that the contribution of linguistics to literary study is minimal; there have been some important studies at the level of phonology and phonetics that allow us to better understand how the sound pattern or system of a language is important in poetry, for instance.

Gestures in literary studies toward other levels of linguistic analysis include the study of the morphosyntactic patterning of language and its study of word formation (how words are put together with the bricks of the sound pattern of the language such as vowels, consonants, stress, prosody—all that belongs to the phonological level) and word combinations (how we form strings of words one after another), semantics (sometimes the study of lexicology is included here) or the study of the meaning system as a lexical pattern (the accumulation of the thesaurus of its lexicon, say), and the pragmatics or the study of the specific usage of the sound system, the word formation system, the syntax, the lexicon and the meaning patterns and how they are used in specific situations, moments, societies. Pragmatics also includes work done in what is called the Theory of Argumentation launched by the contributions of the French linguist Oswald Ducrot. Dan Sperber and Deirdre Wilson's theory of relevance is also a pragmatic theory.

Within the study of pragmatics is the study of certain approaches found in applied linguistics such as discourse analysis. However, this is a rather idiosyncratic methodology that applies to the particular. Much like Barthes's brilliant yet rather idiosyncratic interpretations in *S/Z*, the analysis and the tools of analysis apply to one specimen of narrative fiction.

At what point, then, can linguistics shed light on a formulation of a general theory of narrative fiction? The first step is to see what pertains to linguistics and what does not. For instance, does discourse analysis have anything to do with linguistics, actually? A theory of the sound pattern in English perhaps only identifies a particular instantiation of the universal properties of the faculty of language and therefore cannot operate at the level that would allow it to generate a general theory of language.

Once we make these distinctions, then we have to ask why, for instance, studies of phonetics and phonology have been useful for the study of poetry. Why and in which specific way have the findings of linguistics in the field of phonetics and phonology been applicable to the study of something as particular as poems and something more general but still particular like poetry within such and such a language? And why is this not the case for other fields pertaining to linguistics?

Even if we have as our focus a novel like Joyce's *Finnegans Wake* I am not sure you need a huge theory of morphosyntactics, semantics, or pragmatics to enrich an understanding of its aesthetics. (I speak to this more in chapter 4.) Precisely because it is not just a sum of linguistic devices, it could be that even a narrative fiction that plays on multiple levels with language *does not* benefit much from the use of scientific findings of linguistics.

We have magnificent fictions such as *Absalom! Absalom!* with very long sentences that sometimes run for pages, yet I am not so sure we need the heavy apparatus of linguistic science to describe what Faulkner does here and how it works—to figure out how he can build such long sentences without losing his readers, for instance. Maybe we do not need the heavy apparatus of the science of language to describe and show how this works. Perhaps, too, an ordinary school-level grammar textbook would do just as well.

To put it in a nutshell: the language in narrative fiction is rarely a constant flaunting of linguistic devices, whatever they may be and at whatever level. Narrative fiction rather almost always features the use of everyday ordinary language with the addition of some less frequently used words—words that you find in dictionaries. Maybe the application of linguistics to the study of narrative fiction is simply the ability to use a dictionary. I suppose what I'm saying is that perhaps it is a bit excessive to use the heavy machinery of syntactics and semantics in the study of literature.

PCH: Of course, none of this is to say that literary critics should either vigorously take up current linguistic theories or rely on linguists to extend current theories to literary concerns. After all, even if a particular theory is now widely accepted (and most linguistic theories are highly contested), it is almost certain to be discarded eventually. This may be true particularly in linguistic theory today. As John Ingram points out, "If current trends of technical and scientific advances continue, all currently competitive theories of language processing will probably seem ridiculously simplistic from a vantage point not far into the new millennium" (14). Part of the reason that Labov's insights have been so enduring is that they are not too narrowly tied to a particular, technical theory. Put differently, sometimes the basic descriptive account of some literary phenomena is so interwoven with complex (and doubtful) theoretical presuppositions that it is difficult to extricate the insights from the presuppositions once the theory is discarded. One might argue that this is the case with certain aspects of psychoanalytic work, to take an example from outside of linguistics. For instance, much of Peter Brooks's account of narrative relies on a psychoanalytic account of drives that does not seem to have much validity in terms of current theories of human motivation systems.

At the same time, this does not mean that we should discard current linguistic theories (or current theories in affective science or elsewhere). Rather, it means (I take it) that we should try to learn about the theories in their complex, particular arguments and analyses so that we can approach them critically. At the same time, we should learn about alternative approaches in linguistics-related fields. Finally, we should pay attention to the specificity of verbal art itself, seeking to capture its properties in ways that can be preserved across changes in theoretical fashion.

FLA: Your salient points lead us back to basic distinctions: theoretical versus applied linguistics. It is interesting that you mention Labov as the most enduring example. Perhaps he has endured both because his so-called linguistic work is less dependent on one linguistic theory and because it does not pertain to the field of theoretical linguistics but rather to applied linguistics. Hence, symptomatically, he identifies his own work as sociolinguistics. Indeed, Labov's work easily found its way into schools at all levels from elementary all the way up through high school, giving teachers the necessary knowledge to show that African Americans were not speaking incorrect substandard English. All these findings were very easily appropriated by activists everywhere in favor of civil rights and the civil rights movement. Now it is part and parcel of our ordinary opinions. Labov's findings have become a part of today's common doxa.

This has nothing to do with linguistics as a science or with the study of narrative fiction. One way or another, Labov's work is more an applied sociology than an applied linguistics.

Yes, we have to study very carefully as far as we can the different offerings we have today in the scientific study of language—linguistics as a science—and try as hard as we can to understand the details in order to be able to decide if any of these options in science might be useful for our analysis of literary texts.

The existence of options today in linguistic science is not in any way a negative feature—a lack, say, of a scientific quality of linguistics. In many ways we have the same phenomena in most sciences as with, for example, physics: there are perfectly good scientific reasons to accept the hypothesis of black matter, but at the same time there are very good reasons to reject the hypothesis of black matter, too. The same can be argued and counterargued concerning the ultimate structure of the universe. Some find string theory compelling and believe that the ultimate components of matter are strings; others consider it not valid and on the contrary useless, arguing that it complicates things and keeps people from going deeper into the research.

PCH: Before going on to some particular theoretical approaches, there are a couple of further general points that are worth pursuing. First, we have been speaking of direct applications of linguistic theories to verbal art. Such applications have evident value when literature falls within the scope of the initial theory. However, literary critics have been perhaps overly generous in their interpretation of what falls within the scope of a linguistic theory. Indeed, they sometimes seem to have assumed that the various objects of literary study—for example, narrative—are subcomponents of an overarching language system. But this does not seem very likely. We may wish to view narrative as modular (thus as an autonomous system that "interfaces" with language, etc.) or we may understand it as "emerging" from the interaction of various distinct systems (language, emotion, causal inference, theory of mind, simulation, etc.). In either case, we cannot assume that all the structures, processes, and contents peculiar to language apply directly to narrative. Although, of course, they will operate in the language of narrative, they need not operate in the processing of, say, story structure.

This is all a bit abstract, so let me provide a more tangible example. Suppose we accept for the moment the existence of a language module. That language module will involve certain structures (e.g., perhaps distinct phonology, morphology, and syntax components), certain processes (e.g., in an early generative theory, transformations), and certain contents (e.g., phonemes marked by distinctive features). It does not follow that narrative operates in precisely the same way. In other words, it does not follow that there are distinctive features defining elementary constituents of narrative. Nor does it follow that there are transformations underlying story structure. Much linguistically oriented work in narrative theory is in my view vitiated by sometimes implicit, sometimes explicit assumptions of some identity or continuity along these lines.[1]

FLA: I agree that it doesn't follow that a modular theory equals a theory of narrative fiction. To see this clearly, I think it worth taking a step back to detail the modular hypothesis. The modular theory of mind was formulated in the late 1960s and early 1970s by Jerry Fodor. He has worked on this hypothesis in a systematic way ever since; some have abandoned the hypothesis and others simply stopped writing about it, whereas he has continued to develop the idea. In connection with the modular theory of mind that was very well accepted by Chomsky and rejected by others, he developed the language of thought (LOT) theory—a sort of operating system within the module of language. (See his *Language of Thought*.)

From the beginning it was a controversial hypothesis, and it has been hashed out now for more than thirty years. Today we can say that the hypothesis as such has been clarified and explored in almost all of its aspects. So when we talk about the modular theory of mind and we are for or against the hypothesis, we know what we are talking about. Therefore, it is something manageable that we can talk about in a rational way.

As you well know, the modular hypothesis is that the mind is made up of a whole series of modules—a module of the mind is a region of the brain that specializes, for instance, in language, perception, memory, or movement. By definition each module *is a module* because it operates according to its own principles: the language module is really distinct from the vision module. The problem with the hypothesis is that after a while we have as many modules as we have mental faculties and mental

---

1. See Hogan, "Generative."

activities. Some have multiplied the number of modules in a viral-like way. Like a conductor who helps the musicians in an orchestra play as an organic whole, so too, according to the modular theory of mind, the brain's executive function helps get all the modules to work in a harmonious way in a healthy brain.

The idea of a modular structure of the mind remains controversial, even though concerning certain modules there is more and more empirical confirmation (and therefore less and less debate). For example, research has allowed a mapping within the brain of all the regions involved in language production, and such mapping identifies what we might call a language area, a language module. Some have posited a module for narrative fiction, even.

I find the hypothesis that narrative fiction is a specialized function of a part of the brain highly unconvincing. It does not seem adequate to import every single transformation or operation available in syntax; on the other it is somewhat unappealing to postulate a specific narrative module isolated from other components. I believe we create fiction through the workings of all faculties of the brain. The author of fiction uses all the faculties of her mind. She thinks in terms of causal relations. She thinks in terms of colors, shapes, counterfactual reasoning or imaginings. She counts—that is, she uses mathematical abilities; she uses her linguistic abilities. Her linguistic module of specialization is certainly firing constantly when at work; she is using mental faculties for calculating distances, balance, looking at things without losing sight of what she is supposed to be focused on. Some authors have to walk in order to write, and while walking, they are also using many mental faculties, including the faculty that allows us to shift focus in vision and shift in focus on imagining—these are just some examples of many others.

Quite obviously the creation of narrative fiction makes use of what would be dozens of modules of the brain. We would have to hypothesize that this narrative fiction module would have to be both specialized in fiction and in telling other modules what to do and when to do it at the same time—a kind of super module. In many ways it would be equivalent to the executive brain module.

Yet it is counterintuitive: why would the brain waste energy and time with having two executive brains? Even assuming the modular theory of the brain is true, it is doubtful that one of the modules of the brain would be a narrative fiction module.

Both the production and the consumption or reproduction of narrative fiction involve most if not all the brain faculties, including those faculties that have an inhibitory function that stops us from acting and fleeing a movie theater when the brain signals that what is happening on the screen is not real.

Let us suppose I am wrong and that those who posit the existence of a fiction module are right. Well, this is not a matter of opinion. Today we have the means to determine which areas of the brain become more visibly active when they are playing whatever role they are supposed to play. Through different imaging techniques we know that certain areas are particularly active when we use language. We could use these techniques to explore the brains of writers. It is an empirical question, and we have the means to address this question with all the imaging techniques out there today. So the ball is in the court of those positing the existence of a specialized area of the brain in the creation of fiction to give us empirical proof and to show us if this same area or a different area lights up during the consumption of the fiction.

Further, what kind of distinctly *new* or *innovative* information would we obtain by knowing that it is a module and not the brain working *in toto* or a large number of areas of the brain working together that makes fiction? What would this add to our knowledge? It might tell us something about the functioning of the brain, but does it tell us something about such functioning in relation to the making and consuming of fiction itself? This all speaks yet again to the importance of posing the questions correctly to see the problems and to arrive at an understanding as to how solve them.

PCH: A more recent example of extending linguistic theory to narrative is the use of thematic roles.[2] As Heidi Harley explains, "A thematic role is a general characterization of an argument's role in the situation described by a verb." Thus "an *agent* is an argument that initiates and executes the action of the verb" and a "*patient* is an argument undergoing the verbal action," while "an *experiencer* is an argument whose mental state is affected or described by the verb" (861). Some recent writers have tried to draw on this theory in order to analyze story structures. There are several problems with this. First, we still do not have a decent account of thematic roles or a list that is widely accepted. In other words, the categories themselves are disputed even within linguistics—of the "numerous attempts," Ingram writes, "[n]one have been entirely successful" (34). Second, it is not clear that these categories "scale up" above the sentence

---

2. This is undertaken by David Herman in his influential, wide-ranging, and insightful *Story Logic*.

level. Suppose we have a slave narrative in which we have the following sentences: "John broke his shackles and escaped into the swamp"; "John heard the dogs yelping in the distance"; "After discovering the hideout, the foreman beat John." In the first case, John is an agent; in the second, he is an experiencer; in the third, he is a patient. It does not make any sense to make claims about his thematic role in the story as a whole.

This is not to say that thematic role analysis cannot yield insight into narrative. It can. But it can do so only when applied literally, only when we look at the sentence level and consider whether there are patterns to thematic role assignment there. In other words, here too it is a matter of direct application of linguistic theory. Consider the following passage from late in the 1910 chapter of Faulkner's *The Sound and the Fury* (111–12).

> The three quarters began. The first note sounded, measured and tranquil, serenely peremptory, emptying the unhurried silence for the next one and that's it if people could only change one another forever that way merge like a flame swirling up for an instant then blown cleanly out along the cool eternal dark instead of lying there trying not to think of the swing until all cedars came to have that vivid dead smell of perfume that Benjy hated so. Just by imagining the clump it seemed to me that I could hear whispers secret surges smell the beating of hot blood under wild unsecret flesh watching against red eyelids the swine untethered in pairs rushing coupled into the sea and he we must just stay awake and see evil done for a little while its not always.

The passage is remarkable for many things. One of them comes to light when we think about thematic roles. Human agency is almost entirely occluded in this passage. Quentin is reduced almost entirely to an observer. Sound or time becomes an agent "emptying the unhurried silence." Animals too have agency—thus the swine "rush . . . into the sea." But death, the flame blown out, is in the passive voice with no agent. The most agency that Quentin shows is vainly trying not to be the patient recipient of thoughts—and even there the subject pronoun ("I") is absent. His imagination of the clump is immediately construed as a passive experience that "seemed to me." Other than that, he merely hears and smells. Note that this is a pattern of specific sentences. Indeed, that is what makes the pattern in thematic roles striking. In the story itself, Quentin is very much an agent. Indeed, he is an agent comparable to the swine, for he, like them, is about to plunge to his death in the water. The confinement of his thematic roles in the sentences here is significant precisely because it is not a simple function of some putative thematic role in the story. Throughout the passage, in some way, Quentin is in effect repudiating his own agency, understanding himself as a passive observer of his own death. The point may suggest an emotional numbing that accompanies something like a depressive state.

FLA: With your own superb analysis of the example you invent and the example from Faulkner you are very clearly and swiftly giving definite proof to the fact that the direct transposition of linguistic concepts, hypotheses, and theories into the study of fiction is a mission impossible. Why? Because the initial and fundamental assumption that narrative fiction is structured as a language is simply wrong. Indeed, just because language is used in narrative fiction does not mean linguistics should automatically become the privileged instrument of narrative analysis.

The way you analyze the concept of thematic role shows clearly that this concept is way insufficient to explain even some of the most elementary and simple forms of narrative fiction; it is completely inadequate for explaining the more complex forms, of course. That is, whether we take your invented example or the passage from Faulkner, it is clear that the direct transposition of linguistic concepts in the analysis of narrative fiction can be a dangerous enterprise.

You use the example of thematic role. Of course, many other concepts used in linguistic science appear one way or another in these attempts of what in fact boils down to a *will to replace* a science of narrative fiction with a science of language—when language is only a small part of narrative fiction. If we look statistically at the presence of language in most films as compared to the percentage of images conveying the story, we find that it is much smaller (approximately 30 percent). And then there are films like Spielberg's 1971 *Duel,* which has almost no dialogue, only the terrorizing sound of the tractor-trailer chasing a guy traveling across the desert. Let us not forget either the whole era of silent cinema, where the written word is the occasional signboard appearing on the screen.

This said, and to return to the Faulkner example, the use of linguistic concepts with great precision can perhaps enrich our understanding of how he might select linguistic devices in the making of a nonagentive stream of consciousness we identify with Quentin and that amplify our sense of his passivity. Perhaps, we might argue, that it is only when we know about the theory of thematic roles that we can ultimately understand how Faulkner manages to use language selectively to do so.

However, all this speaks to a larger issue we have been seeing all along: ascertaining what the questions are and what problems we are trying to solve. If one is going to use linguistic concepts and formulations to analyze literature, then one must be clear at the outset what questions one is trying to answer.

There was this argument in the Middle Ages exemplified famously by the question of how many angels can dance on the tip of the needle. This presupposes first that there are angels and second that they can be enumerated and therefore that angels are each one separate, distinctive entities. Last, it presupposes that angels, because of all the aforementioned features, of course occupy a certain amount of space. If we say one or one hundred angels, then we presuppose that they occupy space like any *res extensa*. The question involves a whole ontology of unverifiable entities—there is no experiment, no way we can identify the existence of angels; no way to measure the corporality of an angel. The whole question as an empirical question is meaningless. It refers to a completely invented ontology, and so whatever answer we give to the question is as meaningless or senseless as the question itself. It could be the same if we were to establish an equation between language and narrative fiction, because this assumes a state of affairs—an ontology—in which language (imagine a Venn diagram) covers the whole field of narrative fiction. So whatever questions are asked on the basis of these ontological presuppositions would be as faulty as the ontological presuppositions themselves.

PCH: One way out of the problems we have been considering is to take up features of language that seem to be well established across theories and, even more important, across cognitive systems. Thematic role analysis occurs across competing theories in linguistics and is therefore promising as a method of analysis, when it is applied within its proper linguistic scope. Similarly, at least in my view, Lévi-Strauss's work on transformation sets (in the volumes of *Mythologiques*) manages to avoid the problems of much structuralism. This is because he takes up larger mapping relations. Specifically, it seems likely that a range of cognitive systems involve processes that map structures onto one another in complex ways that are sensitive to context. This is just what Lévi-Strauss finds in myths. It is very different from taking specific principles of transformational displacement from Chomsky's *Syntactic Structures* and extending them to narrative. At the same time, this attention to generalized mapping relations has close connections with some aspects of Chomsky's transformational grammar. (Actually, I would make similar claims for Lacan, whom I read very differently from most Anglophone Lacanians.[3])

Other potentially generalizable features might include the relation between hierarchical and linear structure, or recursion and embedding. Consider recursion and embedding. These are fundamental properties of language. But they also operate in other cognitive systems. The most obvious instance of this is in the recursive embedding of narrators, where Jones begins a story, which includes Smith telling a story about Doe, who tells a story, and so on. This sort of recursive embedding appears from early on, reaching dizzying levels in a work such as the Sanskrit *Kathāsaritsāgara*.

FLA: Yes, at a certain level we can consider how generalizable features of language and cognition might find expression in narrative fiction. I wonder, however, what it would mean to consider narrative fiction as a product of the work of a human being—a human being working (hopefully, I say, because it is statistically probable in a minority of cases) with aesthetic goals in mind. You as author want to tell a story but also engage your reader, so you as author in a very deliberate way make a myriad of choices at all levels of the telling of the story.

The two basic coordinates (elements) of narrative fiction—story and discourse—open the whole field of narrative to aesthetic choice. The choice of a story is already an aesthetic choice, and the way the author chooses to tell the story (picking among the many devices that are possible to use) is an aesthetic choice—of course, in serious writers and not copycat rhetors that use devices mechanically with no aesthetic intent and only because they are useful at getting the message across.

The two examples (invented and Faulkner) that you provide concerning thematic role speak to *deliberate*, conscious efforts at attaining an aesthetic goal. That's why Faulkner, for example, chooses to tell this part of the story in the way that he does; he works on the notions of thematic role and agent and even agency and sets them up in the background until they nearly disappear, making agents out of entities that are not normally presented as such. This is the aesthetic ambition of Faulkner in this case, but it could be a filmmaker, comic-book author/artist, and so on.

Your example of using the concept of the "thematic role" as an analytical tool in an interpretation of the passage from *The Sound and the Fury* (1929) clearly shows that we cannot map linguistic concepts onto narrative fiction. This also holds for all other linguistic concepts.

3. See, for example, "Structure."

Perhaps, however, we are just adding an unnecessary complication by even trying to apply linguistics to narrative fiction?

There have been attempts at using linguistic findings, concepts—even whole linguistic theories—to replace a strict theory of narrative fiction. These attempts have borrowed concepts developed in semiotics (or semiology, according to scholarly preference) on the grounds that since fiction is a matter of signs, the study of signs will allow us to develop fully a theory of fiction. Others have tried to provide a version of psychoanalysis that could become at the same time a theory of narrative fiction. The methods and concepts Lévi-Strauss uses in his analysis of myths have been considered as a possible substitute for a theory of fiction. (His study of myths is meant to tell us something verifiable about the human mind in its most innate form; as such, myths are elevated to the truth correspondence of the factual document. Just as the actual study of myth disappears with Lévi-Strauss, so too does the actual study of narrative fiction disappear with those who treat it also as a factual document of you-fill-in-the-blank.)

That is, the attempts to supplant or replace a proper theory of fiction have been numerous and have come from different quarters (semiotics, psychoanalysis, analysis of myths, etc.). I would not be surprised if quite soon the concepts and the methods used in neurobiology will be the new science that is meant to single-handedly replace a true, proper science of narrative fiction; some in the English departments will read neurobiology and find it compelling and give up narratology entirely and declare that what we have to study is not the text but the human mind and therefore will try to replace the proper study of narrative fiction—the product of human work—with the study of the mind.

I am very mindful of this slippery slope whereby we forget that narrative fiction and the study of the human mind have their own questions, methods, and approaches. Neurobiology can tell me something very interesting about how the mind works in making the aesthetic choices that the duality of story and discourse make possible—a duality that generates the whole field of narrative fiction—and about how it works in our co-construction and consumption of narrative fiction.

Physics cannot replace chemistry and chemistry cannot replace biology *and* linguistics cannot replace narrative theory. Each can explain a lot of the other, but we cannot replace one with the other.

I have not seen this problem posed in the clear way we are posing it because the science of narrative fiction is still very undeveloped. Indeed, it appears that it is an impatient attempt to make a big leap in the understanding of narrative fiction that moves the most level-headed of literary scholars to try to replace this incipient science with other much more developed sciences, including linguistics.

CHOMSKY AND INTERNALISM

PCH: Having mentioned Chomsky, we might now go into his work in a little more detail. There was a brief period of enthusiasm for generative grammar in narratology. However, this faded fairly quickly. For decades, Saussurean linguistics remained dominant, either in its celebratory structuralist form or in its skeptical deconstructive form. Recently, some writers in literary and narrative theory have evidenced enthusiasm for cognitive linguistics and related trends. In most of this, Chomskyan generativist theories remain largely unread—which does not prevent many critics and theorists from dismissing Chomsky on the basis of rumor. Though globally esteemed for his work in linguistics and the philosophy of language, as well as related areas, Chomsky is largely misunderstood and undervalued in literary study.

FLA: I agree wholeheartedly—and wonder with puzzlement why Chomsky has been swept under the carpet, Patrick. Again, I think it might be useful to retrace some of the territory opened up by Saussure then Chomsky. We might then ask if and how they might be useful to those of us interested in enriching our understanding of narrative fiction.

Saussure's ambition was to bring to bear to the study of language as a whole the precision and scientific outlook that the New Grammarians were applying to a diversity of isolated linguistic phenomena. As I have already mentioned, Saussure sought a global vision of language as such and wanted to build a science that would encompass this totality *in a rigorous way*.

He had already begun to train in the New Grammarian tradition as a teenager, but then in Paris he became ill at ease with linguistics as a science, coming to the conclusion that linguists were doing good work but keeping it in the dark because they did not see the whole, only the parts. When he returned to Geneva, this discomfort was expressed in the attempts he made in the courses he taught at the university to clarify two important elements: the proper territory of linguistics and the most appropriate methods for studying this delimited territory. Even though he never wrote the book on the subject, the notes compiled by his students and the diligent and creative use of the notes by the editors of the *Cours* allowed those lessons to become an influential book. It was the first and *only* book that explicitly stated that we have to first establish the cartography of the territory and then explicitly establish the tools. That is, it was the first book to establish the methods scholars could use to explore this territory.

The whole of Chomsky's initial work was likewise focused on exactly the same aims. That is, from the huge dissertation up to the small booklet that he published as a spinoff from the dissertation titled *Syntactic Structures* (1957) that put him on the map, he declared explicitly that we do not know what language is nor what linguistics does, nor do we understand what the proper means are of knowing what language is and what linguistics does.

Chomsky uses a very sophisticated logical and mathematical apparatus to give the cartography to the territory: The more explicit it is, the more we can give a clear, mathematical expression to the approach, the easier it will be to confirm or disconfirm whatever is being posited. The first task is to know what we are doing. In order to know what we are doing, we have to draw the borders of the discipline. When Chomsky was studying linguistics as a young teenager, linguistics was already a real mixed bag; all kinds of studies were being fit into the discipline—that which we would classify more or less as variants of applied linguistics. A heck of a lot of what is called descriptive linguistics (the accumulation of a lot of data unaccompanied by an explanation of what the data might mean) was going on at the time.

If we identify explicitly what linguistics is, then we can actually build knowledge and decide on empirical criteria to determine which theory is good. Chomsky gave us the criteria to compare, reject, or accept different theories. Therefore, he gave linguistics a much firmer empirical, scientific foundation than it ever had before.

This is what generated all sorts of enthusiasm. Chomsky had arrived to clean up the mess and supplied the tools for the cleanup—for the discarding of whatever rubbish was being peddled as part of linguistic science. This has been the general achievement of Chomsky since early in his professional life, even from the time he was a student: the delimitation of a territory and the means for exploring and cleaning it up.

His other huge achievement is his placing the domain of linguistics squarely within the functioning of the mind. He considers that the language faculty is a mind/brain faculty; it is an area of the brain. Therefore, at the same time that he gives a deeply mentalistic turn to the study of linguistics, he grounds this turn in neurobiology.

If there is anything we can take from Saussure and Chomsky (never usually mentioned in the same breath, but here I do) is that with different means at their disposal, they both set themselves the goal of turning linguistics into a real science of language. Therefore, they directed a lot of intellectual energy to tracing the cartography of the field and to devising the most cutting-edge methods of scientific inquiry to use as tools to explore this delimited field.

Of course, if Saussure's or Chomsky's work and results were to have any application in the domain of narrative fiction, it would be accidental. Happy accidents occur; I am not denying this a priori. It was, however, never either Saussure's or Chomsky's purpose to make any contribution to understanding of narrative fiction.

What those like Gérard Genette, Seymour Chatman, Gerald Prince, David Herman, James Phelan, yourself, and others not mentioned share with Saussure and Chomsky is a scientific worldview. In a certain way, by working and proceeding the way he did, Chomsky energized a lot of people and made them confident that the study of mind could be a scientific endeavor, contrary to what behaviorists claimed. The task at hand was to precisely delimit the territory and devise the necessary tools. Those who went this way made real contributions to science and our understanding of reality. We ask, what delimits the territory of fiction and poetry? What do we need to do to delimit and to explore this territory? So our real problem is to determine the field and the tools for exploring this field of narrative fiction.

PCH: As you know, Chomsky-like or "generative" approaches to literature have been displaced almost entirely by cognitive linguistic approaches. I certainly see value in cognitive linguistic study of literature. For example, I believe that Lakoff and Turner's *More than Cool Reason* is a brilliant book. As is clear in these conversations, I have significant disagreements with some cognitive linguistic claims as well. But, for present purposes, my most profound disquiet with the state of literature and linguistics has to do with the ways in which Chomsky and other theorists are misrepresented and dismissed.

Obviously, time does not allow us to consider much of Chomsky's work. However, we might touch on a couple of points, particularly some recent theoretical developments in generativism. We might begin with a few basic architectural features of Chomsky's recent account of language. There is a language module and thus a cognitively autonomous language system. That language module can produce infinitely many well-formed sentences. Precisely because it is autonomous, the language module must interface with other cognitive systems. Chomsky identifies two such systems: the sensory-motor system and the conceptual-intentional system. The development of Chomsky's current ideas, termed "the Minimalist Program," involves considerations of computational complexity in the language module (including the elimination of separate phrase structure and transformational components [see Lasnik and citations]). It also involves an analysis of the degree to which the interfaces with other modules are or are not optimal. A key point here is that the language module produces hierarchical structures. Consider, for example, the sentence "Who does Quimby say ate the last Twinkie?" We connect "who" with its structural partner "ate," not with its linear partner "does" (Quimby being the one who "does . . . say"). This hierarchical organization is a ubiquitous feature of language. Hierarchy does not cause any problems for the conceptual-intentional system, since it is also viewed as hierarchically structured, according to this account. However, the sensory-motor system is linear. This produces a degree of nonoptimality at the language/sensory-motor interface.

In combination with other factors, the relative optimality of the language/conceptual-intentional interface and relative nonoptimality of the language/sensory-motor interface suggest something about the nature of language. Specifically, in Chomsky's view, language evolved not as a communication system but as a system of thought. Had it developed as a communication system, then we would expect greater optimality at the sensory-motor interface, since that is what allows for expressing or receiving communication. Chomsky bolsters his argument by pointing out that mutations do not take place in groups. They take place in individuals. Whatever mutations produced the language module, they had to occur in individuals. As such, they would be useless for communication, since no one else would have them yet. If language-generating mutations had selective advantages, those advantages must have been a matter of thought, at least initially.

I am not entirely convinced that Chomsky is right about this (as I will explain in a moment). However, it seems clear that he is providing a valuable counterweight to a widespread, unconsidered prejudice in the study of language. This is the prejudice that language is simply communicative. In fact, it seems that the great majority of language use is not communicative at all. The great majority of language use seems indeed to be a matter of thought. But this is almost entirely ignored. For example, discourse analysis is defined as the study of language above the level of the sentence. However, it is almost invariably treated as virtually identical with pragmatics, which concerns speaker–hearer interaction, despite the fact that a great portion of our discourse must be internal.

The point is relevant to literature—and particularly salient in Modernist literature—which frequently represents noncommunicative speech in the form of directly or indirectly presented verbal thought. Of course, this is not to say that the *author* is using this speech noncommunicatively. He or she is definitely communicating—an important and consequential point, as theorists such as James Phelan have shown. Nonetheless, in the storyworld, inner speech is not communicative, at least not directly.

FLA: If I understand you correctly, it would seem that a theory of literature might also contribute to the study of internalism: in literature language is used to represents thoughts or other levels of consciousness.

PCH: I wouldn't say "for the study of internalism," but "for the study of language and thought," as I hardly want to commit myself to an internalist program.

FLA: I think this is a very interesting idea, and it points to a potential expansion, at another time and in another venue, of Fodor's Language of Thought idea. Let me follow another train of thought here. Since one way or another language and thought are connected with action in any of its guises, it is clear that language has to connect with a system that involves action (some form of activity, form of behavior). I can give public expression to my thoughts through the phonetic system that involves the movement of a whole series of muscles and nerves (mouth, tongue, etc.), and this is just one among several sensory-motor systems. And the use of the language faculty sets in motion the mental faculties associated with goal setting and deliberately oriented (focused) thought. The intentional part of this bipartite entity is that language implies the application of a focus, an orientation, a goal, and also that it is language connected to the mind/brain (mental) system that gives content to the thought that is expressed or formed through language. So conceptual-intentional is one unit that gives orientation or purpose or goal to language expression or language use. The same is true of concepts. Both refer to the content, shape, and orientation of thought as shaped by the language faculty (as in universal language).

My point here is that you summarize Chomsky's hypothesis very well.

Yes, Chomsky does not believe that the language faculty developed for purposes of communication—the common doxa since Aristotle and Plato. Since Chomsky (who picks up other philosophers' positions on this), very few people today insist that language is not essentially a matter of thought. Today we know that the language faculty forms and gives shape to thought. The communicative function is totally secondary. The trait that defines language is thought rather than communication.

From the point of view of evolution and the presence of language with the birth of homo sapiens, language was born and language developed as a thought-shaping instrument. It is a completely internal instrument. Language could exist and in fact existed at a certain time without any communication between one human being and another.

From this point of view, Chomsky's position is so radical that he considers it likely in our evolution that in individual A, and individual B, and individual C, and so on, each developed this strange capacity that we call the language faculty that boils down to a mechanism that shapes thought. Therefore, it is something completely individual; before having any social function it existed in individuals. But because language made individual A, B, and C more apt for survival, when they reproduced they spread more and more this faculty of language through their genes they were able to transmit to children, and it gave them an advantage (focused thought allows for tool-making advantage and our transformation of nature, etc.) over those without this faculty. So eventually in most or all homo sapiens you found the faculty of language.

As you so pertinently remind us, the common doxa says that language originated in the need of one person to communicate with another. This is what is called the externalist view (vs. Chomsky's internalist view, which is why he names the language faculty the "i" language; this stands in contrast to the "e" language that is English, Spanish, etc., that is used as a vehicle for communication).

What I would like to insist on is that Chomsky's position is a radically immanent position. It is a radically internalist approach. This entails that we cannot mix internalism with externalism. They are radically opposed. Any attempt at fusing or adopting an eclectic position—partly internalist and partly externalist—does not make sense. Chomsky's radical internalism has a very specific theoretical and therefore explanatory function within his theory.

Either you go the internalist way (Chomsky) or the externalist way, and the implications are radically different, including for our understanding of evolution. This is why eclecticism here does not work. This is an empirical matter. We can try to carry out large-scale studies of what happens in people's minds, say, during the whole time they are awake and see if statistically we spend more time communicating together with people or if we spend more time with our own internal thoughts.

This is also an empirical question; we could design such a study with relative ease. Beyond this study that would not require a huge amount of technology, we can think about our own everyday activities. I am in constant interaction with my daughter Corina, for instance. Even though I am with her many hours a day, the part of my thinking that is verbalized is still a small part. Why? I am not verbalizing the moment when I step around a computer cable that runs from the kitchen table to the socket. I am not verbalizing the thought that I should get a better backpack for Corina. I am not verbalizing the fact that I need to use the bathroom. I am not verbalizing that I'd like to sit down. Examples detailing what is going on in my mind compared with what I actually communicated can be multiplied ad infinitum. And this is true of everybody during the eighteen or so hours they are awake. Most of the linguistic shaping of my thought takes place internally with no communicative intention. Again, we do not have to take my word for it. It is perfectly possible to devise a study that would gather data from a lot of people for statistically significant results.

I think Chomsky is right that the common doxa is wrong and therefore so is the approach. Language is not essentially a communication device, and therefore the concepts and hypotheses derived from this empirically and theoretically incorrect position and approach lead to a whole series of misunderstandings concerning the nature of the language faculty as such and of the functioning of this language faculty.

LITERATURE, INTERNALISM, AND DIALOGUE

PCH: Continuing with Chomsky's internalism for a moment, I would say that, for researchers in literature, even a very weak version of internalism has several apparent implications. First and most obviously, as I noted briefly before, it suggests that the representation of internal, linguistic thought in literature should be highly prized as a source of understanding language. To a great extent, linguistics has focused on language that is "externalized," spoken, written, or signed by an utterer to an addressee. Of course, authors are externalizing when they represent internal thought. However, they are doing so with the particular aim of depicting unuttered language. Moreover, readers respond to those depictions in ways that suggest their emotional power, if nothing else. Needless to say, authors can get things wrong. But the success of many authors' depictions of unuttered speech suggests that their depictions are not entirely fanciful. Each of us has experience of our own unspoken language. It seems unlikely that an author's representation of such speech would affect us if it were wholly inaccurate.

Put simply, Chomsky's internalism suggests that literature should have a role in the development of linguistics, and not simply the reverse. Moreover, just as the contribution of linguistics to literature has been largely in the hands of linguists, this contribution of literature to linguistics should be largely in the hands of literary critics and theorists. I say this because the understanding of unuttered speech in literature requires nuanced hermeneutic analysis of the sort that literary critics are trained in. Specifically, even at its most fully developed—even in the form of interior monologue and stream of consciousness—literature cannot simply present a transcription of unuttered speech. There seem to be at least three ways in which literary depictions are likely to deviate from unuttered speech and thus three situations in which critical and interpretive evaluation are needed.

First, writers can get things wrong in ways that are emotionally effective. While we would not expect entirely fanciful depictions to be effective, we would expect certain sorts of nonmimetic developments to be effective. For example, we would expect depictions of interior speech to be more coherent with the author's aesthetic and thematic goals than would ever occur in real interior speech. Thus we would expect there to be certain sorts of "idealization." Second, all inner speech has to be represented in its sensory-motor form. Thus it has to be translated into the sort of linearity that Chomsky claims does not characterize either the language module or the conceptual-intentional system. Finally, a range of nonlanguage thought must be represented in linguistic form. Thus, for example, perception and attentional orientation probably are not linguistically identified in thought. However, in literature, they need to be signaled, and the only way of doing this is through language. (The problem is partially mitigated in film, graphic fiction, and other partially nonlinguistic outlets, but these commonly have other problems with the depiction of inner speech.)

FLA: I understand well your position here. However, I have found in my work that it is problematic to ascribe to characters the neurobiology and the behavior of real flesh-and-blood people and that it is problematic to ascribe to fiction a mimetic function.

I hold to the notion that the blueprint of narrative fiction is a new product—like the making of a table that adds something new to the world. When Joyce uses the interior monologue to represent the mind of Molly Bloom in *Ulysses* (1922), he's seeking devices to create an aesthetic effect; he's trying to move his reader through a series of completely artificial, completely created sentences, and he's not reproducing any real thought processes in *anybody*. For his own aesthetic purposes he's writing specific sentences that he wishes the reader to read as being at the same time a highly elaborate, highly aestheticized representation of mind processes (which he conveys through the typography and lack of punctuation and lack of segmentivity in the monologue). He wants to dazzle the reader with his virtuosity. He wants us to experience an aesthetic satisfaction through the reading of this chapter. And at the same time, he does not want the devices to be so overwhelming that the whole set of sentences composing the chapter are not felt by the reader to be a recognizable representation of mental processes.

The aesthetic purpose he fixes for himself in *Finnegans Wake* (1939) is to reproduce the functioning of the mind while it is dreaming. Nobody dreams the way the drunken HCE (Humphrey Chimpden Earwicker) dreams in *Finnegans Wake*. I have never dreamed in forms of thought using a morphology where I make words out of fragments of words from a dozen or so languages.

From my side of the table, I see Faulkner's novel as a highly aestheticized version of reality. Like all serious writers of narrative fiction, he uses for his own aesthetic purposes whatever bricks he wants from the world, and his product, the result, is built with aesthetic purposes in mind and in no way imitates or reflects reality.

PCH: I'm not ascribing neurobiology to characters. I'm simply accepting the standard cognitive view that simulation and emotional response do not suddenly change in some radical way with fiction. Since you mention Faulkner, we might return again to the passage from *The Sound and the Fury*. We can see from the start that it combines perceptual and linguistic responses. The first note sounding should probably be understood primarily as a perception accompanied by an implicit anticipation of further notes. This anticipation does not require language since we find the same sort of anticipation in nonhuman animals. Similarly, the attribution of serenity to the bell should probably be understood as a tacit emotional attribution that is largely language independent. Later in the passage, "and he" signals a shift in memory or imagination that is presumably not linguistic, but merely presented as linguistic for the reader. In contrast with these nonverbal experiences, the following statement from Mr. Compson ("we must just stay awake," and so on) is almost certainly subvocalized. In other words, it is almost certainly encoded in the sensory-motor system and rendered linear in inner speech. Indeed, that is probably the case for the entire following dialogue between Quentin and his father, even parts that are imagined rather than remembered. Finally, Quentin's reflections on "the cool eternal dark" and the "vivid dead smell of perfume" may suggest internal thought that is in some way linguistic, yet not subvocalized.

Thus this brief passage points to different sorts of internal experience. All are similarly represented in externalized form. However, there are indications that some are nonlinguistic; others are subvocalized, thus linear; and, perhaps, still others are (in Chomsky's terms) produced by the language faculty but not interfaced with the sensory-motor system in subvocalization.

FLA: I see more clearly now what you are getting at, Patrick. Yes, Faulkner describes a whole series of what we could call internal mental processes that are written in such a way that they are made to imitate internal mental states and activities. He also describes events taking place externally, outside the mind. So we see in this passage a mixture of internalism and externalism—but I wonder if we should add to this that whatever is considered the internal mechanisms of the mind of the protagonist and that which is written outside of his mind *is written by Faulkner.* He deliberately aims to make an artifact that can create the aesthetic effect on the readers of identifying processes in what they are reading that are somewhat akin (or similar) to mental processes with which they are familiar.

We derive a certain aesthetic pleasure in finding our way around Mexico City as it was in the 1950s when we read Carlos Fuentes's 1958 novel *La region mas Transparente* (*Where the Air Is Clear*) for the first time. There is a specific pleasure in recognizing features of a city you might know well. This probably happened to the first readers of *Ulysses.* This is something deliberately sought by Joyce and Fuentes. It is their aesthetic aim at that moment and concerning those specific passages in their writing. But we also derive an intense aesthetic pleasure in the description of Paris when a giant arrives in the city in order to study at the Sorbonne and suddenly feels the urge to pee and does so and floods half the city and drowns all sorts of people and animals that are accounted for in an encyclopedic way.

In other words, Rabelais' Paris is no less fictional than Joyce's Dublin and Fuentes's Mexico City. Just because in one case the author adheres to the aesthetic of the grotesque and in the other cases we see the adherence to the aesthetic of realism does not mean all aren't equally constructs. They are not in any way reproductions or simulacra or imitations of anything.

PCH: We seem to be disagreeing here, though that may be a matter of what you mean by "imitation." In my view, simulation may be more or less constrained by real-world principles, though it is never wholly discontinuous with those principles. Moreover, our response to fictional simulation is not isolated from our response to the real world. Take a film. A character expresses sorrow by weeping (the actor mimics weeping). A viewer's aesthetic response to that is connected with mirroring—and mirroring operates on real-world principles connecting grief with weeping.

It might be worth thinking a little more about the Faulkner passage. We might ask if it really makes sense to say that there is internal speech that is not subvocalized. Clearly, there is something like nonsubvocalized thought (e.g., perception followed by expectation). Moreover, there is evidence that encoding a memory in (external) speech alters the memory (see Anderson, "Incidental" 208–9). This seems to suggest that speech encoding is not automatic and that we can remember events—thus presumably think about them—without thereby using speech, internal or external. But none of this tells us whether there is linguistic, nonsubvocalized speech.

To consider this issue, we might do something reminiscent of Vygotsky and turn the question around and ask to what extent inner speech is social. More precisely, we might ask to what extent inner speech is tacitly directed to an addressee. Speaking for myself, it certainly seems that a lot of my own internal speech is "dialogic" in the sense of being oriented toward someone. Indeed, when compared with characters in works by Joyce, Woolf, or Faulkner, it seems that far more of my own internal speech is directed toward an addressee—perhaps almost all of it.[4]

Of course, this hardly resolves things. It may simply return us to the issue of whether only a part of internal speech is subvocalized. It may simply be that internal dialogic speech is subvocalized and that subvocalized speech is more salient than nonsubvocalized inner speech.

---

4. For further discussion of these topics, see chapter 9 of Hogan, *Ulysses*.

On the other hand, stressing dialogue even in inner speech has some potentially significant consequences in assessing Chomskyan grammatical theory. For example, a central issue in Chomskyan analysis concerns putative "silent" items (formerly called "traces"; see Surányi, "Principles" 669). Consider, again, the sentence "Who did Quimby say ate the last Twinkie?" In a standard generative analysis, there are two instances of "who" in the sentence—one is spoken and the other is not. The first appears at the beginning of the sentence. The second appears between "say" and "ate." This has many consequences that are widely discussed in the technical literature. In the sort of purely interior model adopted by Chomsky, the silent copy results from grammatical operations that serve to move "who" from an initial position between "say" and "ate" (what is referred to as "wh-movement" in linguistics literature). But it can be argued that at least many of the data are adequately explained by placing such questions in the context of dialogue, where there is an implicit answer.[5] The "silent copy" is then not the result of some past, wholly interior movement. Rather, it is the result of a tacitly anticipated response to the question (e.g., "Quimby said *that bastard Waldo* ate the last Twinkie!"). (Not all traces appear in questions. But the point here is not to solve every grammatical problem in a single paragraph. Rather, the point is to suggest possible avenues of inquiry that may be promising but have not been explored.[6])

This is related to another issue. In conversation, we continually engage in self-monitoring, "taking account of the listener's perspective" (Ingram 346; see also 25). This self-monitoring is based on a continual tacit simulation of another person's understanding, thus our tacit simulation of his or her likely response. This fits with the idea that an implicit dialogic response could account for wh-movement. But that link suggests that some apparent grammatical constraints are at least in part a function of monitoring audience-directed speech, including subvocalized speech. This, in turn, may lead us to an extreme "dialogic" account of inner speech. In apparently diametric opposition to the Chomskyan view, this would posit that, in most cases, our internal monologue is tacitly dialogue, that it involves a degree of self-monitoring, some tacit orientation to an addressee, even if the "speaker" is not self-consciously aware of it. (Actually, things are more complicated, since internal dialogue is not actual dialogue. Even if all our inner speech is dialogic in this sense, the primary operation of speech may well be internal, thus more a matter of thought than actual communication.)

5. See Hogan, *On Interpretation* 76.
6. See Hogan, *On Interpretation* 77.

In connection with this, it is worth remarking that self-monitoring arises in recursively embedded simulations, both in ordinary life and in fiction. An author not only simulates a reader. He or she simulates characters. Part of that character simulation involves embedding a simulation of a character simulating other characters and monitoring himself or herself in relation to those other characters.

Here we might turn once more to the Faulkner passage. If we combine extreme dialogism with recursive embedding, we might expect all or almost all linguistic components of the passage to involve audience address and self-monitoring. I should say that I am not actually proposing the extreme dialogic view. I am merely suggesting it as a possibility to be considered, along with Chomsky's account and intermediate options. In this context, we still need to separate out perceptual and attentional concerns, since these would be a matter of the author self-monitoring and adjusting for the reader. However, when we come upon a clearly linguistic passage that does not seem to be oriented to an addressee, we might question whether there is an addressee tacitly simulated. Note that, as the word "tacitly" suggests, this is not something that would necessarily be known to the author. The author would proceed with his or her simulation in the ordinary way, without necessarily having any metacognitive awareness of what precise contents or processes were involved.

In the case of Faulkner, there is a striking change with "that's it." That does not seem to me to be oriented to an addressee. Rather, it is a statement that is used as a spontaneous exclamation of realization. Such ejaculatory utterances probably escape self-monitoring all the time in ordinary conversation. But what follows is different. Quentin thinks, "if people could only change one another forever that way." Before, I took this as, roughly, an expression of Quentin's true inner thought. But, in light of what follows, it makes at least as much sense to read it as an initial statement tacitly addressed to his father. In that context, Quentin's realization is not so much a matter of what he really believes. It is not a moment of self-discovery. It is, rather, a moment when Quentin realizes just what he should have said in his conversation with his father. It is, in that sense, oriented to a recipient—thus it is, for example, potentially just as unreliable, though unreliability is something we do not ordinarily associate with interior monologue.

FLA: You are right to mention Vygotsky, Patrick. He drew important theoretical conclusions from his observation of children; that they naturally accompany their activities with verbal (and song) expression. I know this firsthand, of course, watching Corina play with dolls or draw or put away her toys or clothes: all kinds of sounds, narratives, ascriptions of thought, and so on, accompany her play and activities. It is a universal trait. It is a type of dress rehearsal for the normal regular use of language in older children and adults. That is, it is a dress rehearsal of language *as a tool for communication*. Just like Chomsky is a good example of radical internalism, Vygotsky is an excellent example of externalism.

I am not sure the use of both works. It is convincing to the extent that we cannot observe the thought processes taking place in newborn babies—before they are able to give linguistic shape to their thoughts. It would seem that at a certain point in time, age two or so, when they begin to speak, they use external manifestations of language such as those that Corina uses when she plays with her dolls.

Indeed, Alison Gopnik and others have devised ways to infer the mental processes even of newborns, by, for example, the amount of time their eyesight is fixed on something that is considered novel way before they can utter sentences or even words.

I wonder, too, about thinking in dialogue with an implicit addressee. When I am thinking, I am not in dialogue with someone. I am not doing what, say, Corina does when she's playing. Sometimes I am not even sure what language I am using when I think. Thinking can be enjoyable and also a painful activity. It can be to enter a labyrinth where I get lost and find it difficult to focus and whatever I focus on is not what I want to focus on. It can be painful in the sense of it being a difficult activity.

In any case, whether thinking is dialogical or not or whether it implies the implicit evocation of an addressee or not, the fact remains that thinking in narrative fiction is a construct. I am interested in fiction for the aesthetic construction of a character like Raskolnikov's capacity to reason and other mental processes.

In generative grammar the description of language and its functioning uses concepts and operations that refer to processes that we are not normally aware of. They are concepts and processes discovered by science. There is the law of gravity. You have to be someone schooled in the concepts to be able to talk about the law of gravity and concept implied in the law of gravity. When I walk or ride my bike, I know nothing about the law of gravity, yet I still walk and use my bike. It is the same with use of language. I use my language, but I am not aware of hidden concepts and mechanisms/rules of operation of the faculty of language.

Like all regularities, all laws of nature, all scientific discoveries and facts explained by science, we seldom know or are familiar with or are aware of the theories behind these sciences. Science is almost by definition a science of that which is not immediately perceptible. It is hypotheses, theories, and concepts that explain the functioning of reality that is not immediately and directly perceptible.

CONNECTIONISM

PCH: Before going on, a brief terminological point is perhaps worth making, primarily for the benefit of readers. "Externalism" technically refers to the view that meaning must involve nonpsychological elements. For example, suppose the meaning of "water" in my mental lexicon is "stuff like this" (as I point to the clear liquid that is coming out of my tap). Then "water" means whatever that stuff is. Since that stuff is H_2O, then water is H_2O. But it does not matter if anyone knows the chemical composition of water. That is still what water is, and thus it is what "water" means. (On externalism, see Deutsch and citations.) This makes meaning in part contingent on nonpsychological facts. I think this is a mistaken view. But, whatever one thinks, it is different from the view I am presenting. In the view I have just suggested, monologue is oriented toward an addressee, even when it is all going on subvocally (i.e., without utterance). Thus the meaning is still all "internal" in the internalist versus externalist sense.

I want to stress this because the ideas are so often mixed up. There is a common view that externalism and social orientation—and even dialogue—all go together. A recent example is Alan Palmer. He in effect takes up the old conduit metaphor (see Reddy) that some sort of meaning enters into signals and then the recipient of the signals takes out the meaning. This is presumably what makes "much of our thought . . . *visible*" (197, italics in the original) rather than merely open to inference. This visibility is, it would seem, what supposedly allows "intermental thought," as posited by Palmer. Palmer associates this intermental thought with philosophical externalism, dialogism, and many other things. The basic premise of Palmer's view does not seem at all plausible to me. In fact, it seems to be an instance of the common cognitive bias—the "illusion of transparency" (see Keysar and Barr 158-63)—that leads people to view communicative meanings as obvious, even when they are opaque or highly ambiguous. Palmer's expansion of this tendency does have a noble lineage. It is basically an idealist view, reminiscent of Hegel. The key point, however, is that it goes well beyond philosophical externalism, which simply makes meaning contingent on certain physical facts.

Clearly, there is much more to say about generativism, internalism, and externalism. However, given constraints of space, we might at this point turn to another important approach to language and the mind, connectionism—also called parallel distributed processing (PDP) or neural network theory—one of the major alternatives to generativism.[7] Connectionism understands cognitive processes to be based on a simplified architecture of the same general sort that we find in the brain. Thus there are neuronlike units, which have various degrees of activation and which are connected to one another with excitatory or inhibitory links in various degrees of strength. Cognitive processes are understood entirely in terms of circuits of activation. These processes operate in parallel with one another, and their constituents are distributed across different units.

For example, in a connectionist account of an emotional response to an event, we might find something of the following sort. A number of different, distributed "probes" would produce different degrees of activation in a number of different memories. These would in turn have connections with one another and with different emotion systems. The emotion systems might have inhibitory or excitatory relations with one another. The final state of the system would involve some activated memories and some emotional response. Suppose, for instance, that Smith has a series of memories bearing on his parents' home. Entering that home just after the death of a parent will trigger an array of different memories. Some of these memories will be stronger (i.e., have a higher activation level) than others. Some will be happy; others will be angry; still others (such as the memory of the recent death) will be sad. In other words, different memories will activate different emotion systems. These systems are not mutually compatible, so that they are likely to have inhibitory effects on one another. The result may be that Jones experiences shifts among the emotions—sadness for a moment, joy in recollection that displaces the sadness, grief that then returns, and so on. These waves of feeling may be understood as the network-based operation of different patterns of activation and inhibition producing different (ephemeral) mnemonic and emotional configurations.

FLA: I see the attraction to the way connectionism works according to 1/0 algorithms. It is a sequential model (sequenced in parallel, which gives it much more computational power), and there is an analogy between

---

7. For an introduction to connectionism, see Hogan, *Cognitive Science* 48–58; for fuller treatments, see Dawson and McLeod, Plunkett, and Rolls.

this procedure and the way written and even spoken language appears—always spatially sequential and linear, beginning from left to right or from right to left in written form. It stretches in space and it stretches in time in a linear way. Even orally, language works sequentially, one word after another after another. And it always takes place in a linear time pattern. Therefore, there is a strong analogy with language with the way connectionism works.

In the connectionist view, every operation in the mind/brain is a connectionist operation—one that follows computer-like algorithms. Today's multi-core computers all have the capacity for parallel and simultaneously working circuits and algorithms that allow for much more powerful computing power. This was anticipated by those doing work on connectionist approaches. They considered that the fastest computers were in the day clumsy compared with the mind/brain and posited the working in parallel of multiple programs.

Connectionism is a theory of mind that tries to describe (more than explain) all mental phenomena in terms of these sort of algorithms. Of course, the hardware of all these programs, the algorithm, is always material, the brain's neurons—and all neurons function the way the computer functions: open and closed, 0 and 1, neuron active or not active. Emotions, thoughts, language usage—everything mental—in this connectionist view comes down to the operation of neurons in terms of binaries: open/closed, active/inactive, and so on.

Chomskyan generativism is not a general theory of mind. It is a theory of the language faculty and not the workings of the whole mind. Chomsky's aim since the 1950s has been to determine what language is. How can we speak scientifically about language? How can we speak about language that is shapeless and chaotic? How can we study language if we do not first determine what language is? How do we determine the cartography? How do we build the tools to explore this cartography?

Chomsky is first and foremost a linguist. His linguistics includes a theory of mind in which mind/brain is formed by a series of modules. There is a language faculty module, as well as other modules that interface with the language faculty module: 1) the sensory-motor interface; 2) the conceptual-intentional interface. But he does not say that besides these modules that he believes *really exist* in the context of scientific work there is also a module for bike riding or for baking cakes or for making chairs, and so on. He has a theory of what makes the mind/brain, but he does not have a theory that applies universally to all human activities.

Connectionism, on the other hand, wants to tell us that everything we do, feel, all that is happening in the mind/brain, is connectionist. Once we adopt the model of mind/brain as functioning like a computer, then all has to function exactly the same way. I understand well that connectionists and those in the field of statistical language parsing argue that with sophisticated statistical tools and sufficiently large corpora one can actually approximate the generative capacity of phrase structure grammars. However, I am with Chomsky in that I would say that these results are not abstract enough to say something meaningful about the language faculty. In this sense, I am not sure we can compare generativism and connectionism. They do not seem to function at the same explanatory level. The way I see it, we might return to the main question: what is the problem, and how do I solve it?

With respect to the Faulkner passage, what is the problem we are trying to solve with these tools or the mixture of these tools? Do we want to find a one-to-one correspondence between computation theory on the one hand and linguistic theory on the other hand in narrative fiction? Do we want to solve one very particularized problem in a small part of Faulkner's text—a small drop in the whole ocean of narrative fiction? Do we want to see if by solving this problem we will be able to better grasp how narrative fiction works?

The elegant analysis of Faulkner that you provide is a patient, perceptive, careful close reading of the paragraph. And I see how a generativist approach can be useful as a model that seeks to get at foundations. In my experience, however, a one-to-one mapping of a linguistic formulation (generativist or connectionist) onto a theory of literature has not led me very far.

PCH: Evidently, I was not clear. My intention was to say that we should not simply apply a technical linguistic theory to narrative, assuming that the structures, processes, and contents of language define the structures, processes, and contents of narrative. Indeed, I intended to oppose that directly. So there was no question of the reading of Faulkner being "generativist." Rather, the task I proposed was drawing on general features of cognition for general narrative principles and on specific linguistic features for linguistic aspects of narrative. Moreover, this should go in both directions, with literature contributing to our analysis of cognition and language as well as the reverse. The preceding analyses of Faulkner take up the issues of linguistic internality, seeking to explore its nature and varieties, considering, for example, whether there is a difference between subvocalized and nonsubvocalized or addressee-oriented and nonaddressee-oriented interior verbal thought.

Returning to the theories we have been discussing, I would note that although connectionist and generative accounts are considered diametrically opposed, they need not be. Specifically, generative accounts are commonly understood to involve rule-based processing of single objects in serial operations. For example, we have an initial version of a sentence, which is then subjected to merge operations (see Surányi, "Merge"). Needless to say, one need not accept this particular account of a rule-based system. However, it seems clear that we will need to be able to treat cognitive phenomena in both ways. Specifically, something like a PDP approach almost certainly has to be the case since, as far as we know, the brain provides the substrate for all cognitive operations, and neurons appear to operate in a parallel and distributed fashion. On the other hand, it seems clear that languages do follow rules, and we would lose important generalizations if we excluded rule-based processing from our account. For example, we need to be able to account for English plural formation by some sort of neural network. But at the same time we need to recognize that regular plurals are formed by adding "əz" to sibilants (bush/bushes), "z" to voiced nonsibilants, and "s" to unvoiced nonsibilants. In other words, plural formation conforms to a rule. Abandoning such generalizations would be comparable to abandoning the laws of physics in order to give a more fine-grained account of the particulars of causal processes.

On the other hand, it does seem to be the case that thinking in terms of serial, rule-governed processes orients our attention and inferences differently than thinking in terms of PDP systems does. Moreover, it appears that one or the other orientation is more appropriate for any given type of cognitive process. Specifically, we may think of cognitive processing generally as involving two strata. There is an initial generation that is relatively spontaneous. This is then modified when some trigger (usually associated with anterior cingulate cortex; see Ito and colleagues 199) indicates some problem or task conflict—for example, a discrepancy between our aims in speaking and the apparent or simulated understanding of an addressee. The initial generation has a more clearly parallel and distributed character. In contrast, the task-specific modulation involved in self-correction has a more clearly serial and localized operation. In this sense, it makes sense to take up a PDP model in considering the former and a serial model in addressing the latter (even though, ultimately, the two should be mutually translatable).

Here we might return again to Faulkner. Note that the spontaneous, PDP-based production would range over everything from memories to emotions to beliefs to details of phrasing. Only a limited part of this would be affected by monitoring and correction. Here, we might begin with a hypothesis. As we know, Quentin is about to commit suicide. But he hardly ever indicates this directly. If it is correct that suicide is an overarching concern for him (and how could it fail to be?), we would expect it to have widespread consequences in his internal speech, ranging from memories to word choice. However, we might simultaneously conjecture that he could locally suppress direct thoughts of suicide, perhaps as part of orientation to an implicit addressee. (See Anderson, "Motivated," on the suppression of thoughts in a cognitive context.)

Considered in light of a PDP model, Quentin's speech does indeed seem pervaded by thoughts of his suicide, though these thoughts are also disfigured by localized suppressions. First, we have a number of images that suggest partial activation from Quentin's simulation of his death at the bottom of the water. There may be a hint of this in "unhurried silence." There is almost certainly an element of it in the waterlike "swirling" image that gives way to "cool eternal dark." The first and third images may indicate an imaging of death as peace, as a release from the mental torment that Quentin experiences when thinking about his sister's relations with other men. Indeed, the dark of the seafloor contrasts strikingly with Quentin's imaging of his sister's sexual activities, an imaging that occurs "against red eyelids." This analysis fits with the allusion to the Gadarene swine—possessed by demons, the swine free themselves from the pains of madness by plunging into the water. The suffering of the swine is bound up with the sexual nature of Quentin's torment—not because of the original story, but because of Quentin's own addition (unsurprising from a PDP account) that the swine "coupled."

FLA: Now I see more clearly where you are going with this line of thinking, Patrick. I do consider that an understanding of the general features of cognition can and do enrich our understanding of the making and consuming of fiction. Here I would only add that I'm not sure that at the level of its formulation a connectionist view of the mind would be useful. The connectionist approach assumes a rigorous sequential functioning of

0/1, 0/1 binaries. Another way to think of it is when machine-assisted translation (the Internet has plenty of these) seeks a one-to-one correspondence or transportation of syntax pattern A into syntax pattern B. Yet, even with its hugely massive computational capacity in terms of memory, speed, and analogy formation that seeks correspondences that are as close as possible, the translations end up being grab bags of phrases, literally. What I am trying to say is that connectionism might be useful in computer theory, but its overreaching generalization of computers as an explanation of the functioning of the mind/brain generally and in its work to make narrative fiction specifically doesn't seem to work. The generativist model taken at a certain level of abstraction does seem to move us closer to a clearer understanding of the mind/brain's making and consuming of narrative fiction and to an identification of the tools we might use to explore this.

COGNITIVE LINGUISTICS AND EMBODIED COGNITION

PCH: Before concluding, we should probably turn to the approach to language that has been most influential in cognitive literary study—cognitive linguistics. Specifically, it would be worthwhile to consider the value and limits of a recent approach that has given rise to great enthusiasm in a number of fields—so-called embodied cognition.

Unfortunately, "embodied cognition" is used in at least three distinct senses. The first is simply the idea that the brain is the substrate of cognition. For clarity, we might refer to this as "materialism." Basically everyone in cognitive science today agrees on this point. Thus, everyone in the discussion believes in embodied cognition in this sense.

The second sense of "embodied cognition" refers to the guidance of cognition by bodily reactions that are not, so to speak, ratiocinative. For clarity, we might refer to this as "bodily cognition." A prominent example of this is found in mirroring responses, when motor routines are partially activated by observing someone else engage in a certain intentional action or express a certain emotion through facial gestures. At least in part, our mirroring—not some process of self-conscious, logical inference—guides our understanding of someone else's intentions or feelings. This sense of embodied cognition also includes what Damasio refers to as "somatic marking," our use of bodily emotional responses to guide our evaluation of a situation. Cases would include our avoidance of certain sorts of risk to which we have an emotional aversion, even when we do not self-consciously recognize that there is a risk (see Bechara, Damasio, Tranel, and Damasio). Other researchers have isolated similar phenomena under different labels, such as "misattribution." For example, in one well-known case, test subjects experienced physiological arousal due to test conditions but attributed this arousal to the alluring qualities of an interviewer (see Oatley, Keltner, and Jenkins 23, 24). In these cases, the test subjects presumably think that they are making a judgment about the interviewer, when they are being led by their own bodily responses, which have a different source. Damasio's "somatic markers" and misattribution seem to be cases of a more general tendency in human behavior. Specifically, we are driven to act by motivation (or emotion) systems. Part of that motivation involves the isolation of causes and targets of feeling. We tend to assume that we just know these causes and targets. But that is not the case. Rather, we isolate causes and targets fallibly. In the words of Gerald Clore and Andrew Ortony, "People tend to experience their affective feelings as reactions to whatever happens to be in focus at the time" (27). *Bodily cognition* seems to me the most valuable sense of "embodied cognition."

The third sense of *embodied cognition* is the one that is specific to cognitive linguistics. In this usage, our semantics are ultimately derived from our bodily orientation in and engagement with the world. We might call this "body-based cognitive modeling." The basic idea here is that most of our thought is highly metaphorical. Thus we "grasp" concepts; we "move along the path" of life, and so on. But if most of what we think is metaphorical, there is presumably something on which that metaphorical thought is "based." In the view of Lakoff, Johnson, and others, that "basis" is the body. This does not mean simply that the metaphors derive from experience (e.g., perception). Rather, it means that the fundamental organization of thought uses our bodily orientation in the world as a model. (For brief, accessible, and positive treatments of embodied cognition in this sense, see Gibbs, and Ingram 374–79.)

FLA: I agree with you in that there are at least three different meanings of the term "embodied cognition." I think there is a fourth meaning that we find in the work of Chileans Humberto Maturana and Francisco Varela and their disciples, who have been quite influential in France and Latin America: that all cognition is literally *enclosed* within the body—not only in the sense that the body—the brain—is the site of cognition but in the sense that the body as a whole (including the brain once again) determines what *is known* and *can be known* as well as the shape of what we know. In this sense it is a pretty solipsistic variant of idealism.

It is quite evident that cognition is located in the brain and that when we talk about any mind processes we are talking about the mind/brain. So materialism is quite valid, even though I must say there are also variants of materialism that fall short of giving a general explanation for what they purport to explain. Paul and Patricia Churchland posit a connectionist theory of mind to be radically materialist, yet I consider their approach to be much more akin to what certain philosophers call mechanical materialism (or mechanistic materialism or rough materialism).

In the second sense, cognitive responses are also *material reactions* to material realities. In fact, you have in the case of Damasio a theory of emotions (like the one developed by William James) that proposes that something like fear is just the phenomenological manifestation (appearance) of temporally previous alterations, modifications, that have taken place in the body. So we first have the alterations in the body, and then those alterations appear as fear, love, and so forth.

To all this it is imperative to add the fact that the nervous system extends beyond the brain and connects with the automatic nervous system functioning (sympathetic and parasympathetic) of the body.

The third, and most idealistic, sense of embodied cognition is both a highly speculative series of proposals and an idealist worldview: not only our language but also our interactions with the world are determined by the "metaphors we live by"—to use the title of one of Lakoff's books. (See also Lakoff's *Don't Think of an Elephant!* and *Whose Freedom?*) Even from a strictly logical point of view, if all language is metaphorical, then there is no such thing as metaphor. It is like saying that the totality of the universe is gray, which entails that neither gray nor any of the other colors exists. If all is gray, we would not know that it is gray because we only know that something is gray because there is green, blue, red, and so on. If all is metaphorical, then nothing is metaphorical.

So this third sense of embodied cognition is essentially an ideological construct; it has no real basis in reality. This is why it has made absolutely no contribution to the science of language; no scientist worth his or her salt is a cognitive linguist.

PCH: Since the third sense of "embodied cognition" has generated some following in literary study, it perhaps bears further scrutiny. It seems clear that our thought is often guided by metaphorical structures and that these metaphors do often have some basis in our sense of our bodily being in the world. However, one might question the extent to which our thought is guided by metaphors and the degree to which our metaphors derive from body-based cognitive models.

Support for body-based cognitive modeling comes from several sources. The most obvious is the existence of consistent patterns in metaphor. For example, many metaphors have a "source/path/goal" format—as in "I am not making adequate progress in my job." However, empirical research suggests that such metaphors are not ubiquitous and that we do not invariably follow the apparent metaphorical resonances of such (partially concealed) metaphors even when we do use them. For example, corpus linguistic analysis reported by Gerard Steen suggests that only a small minority of lexical units are metaphorical (13.5 percent in the corpora studied) and that almost all of these metaphors (99 percent) are conventional and thus frequently interpreted as categorization. In short, the great majority of speech is not metaphorical, and "a lot of metaphor may not be processed metaphorically" (220).

In keeping with Steen's point about processing, the consistency of ordinary usage is paired with vast inconsistencies in such usage. For example, suppose we are really guided in our thought about a career by the source/path/goal metaphor. This would seem to have a number of consequences. For example, it seems that one should not be able to take a vacation, since one does not typically take a vacation from a journey. Also, "going back to where one began" should entail loss of time, money, and advancement in one's journey. But we often say things like, "I have taken a lot of different paths in my career. Now I am going to take a vacation from theory and go back where I started, to modernism. I think that is where my greatest career success is to be found." In its various parts, a statement of this sort repeatedly draws on a source/path/goal metaphor. But it keeps changing the way in which the target (the speaker's career) is mapped onto the metaphorical source (source/path/goal). The result is that the metaphor is simply not guiding the way we understand the target domain. It is our separate understanding of the target (i.e., the career) that leads to the repeated remappings. In short, our understanding of the target guides our use of the metaphor, not the reverse.

A second source of evidence for body-based cognitive modeling emerges from the body model. There is fMRI research suggesting that word meanings are stored in sensory-motor areas. Thus tool words activate motor areas; animal words activate visual areas, and so on. In fact, as Ingram explains, the fMRI data are actually much less straightforward than this indicates: "Nouns of high imageability . . . may be expected to have more of a posterior (temporal-occipital) locus of representation in the brain, and verbs ('action words'), more of an anterior (frontal) representation. However, conflicting findings have been reported in the imaging literature" (215; see also 216–19 and 233–35). Nonetheless, there is some reason to believe that meanings are connected with relevant sensory-motor cortex. For example, Eysenck cites research that "category knowledge about color, motion, and shape is processed in different regions of the brain, typically in areas close to those associated with processing those kinds of information in visual perception" (124). But this only suggests a certain connection between experience and meaning. It has no consequences for bodily modeling. Indeed, it is not even clear just what the data entail for the relation between experience and meaning. A standard view of semantic memory is that it is built up out of episodic memories (see Baddeley 11), but it becomes increasingly abstract, with particularity fading. For example, we see a lot of cows (pictures of cows, etc.). These are initially stored as episodic memories—largely visual (and partially aural) memories, for those of us who are not dairy farmers. But our semantic memories somehow involve a prototype-like averaging across these instances. If this is in fact the case, then we might expect, for example, semantic memory for visual objects (such as animals) to appear in visual areas associated with visual memories. However, this does not tell us that the semantic memories retain their episodic quality. Such a retaining of episodic quality seems to be the only way in which they might bear on "embodiment."

Finally, embodied cognitive modeling draws evidence from effects of the metaphorical senses of terms. For example, one might argue (in line with the preceding comments on semantic memory) that a word such as "grasp" simply does not have a metaphorical meaning when we say "I couldn't grasp what he was saying." In contrast, Raymond Gibbs claims that "the physical meanings of certain metaphorically used words, like *grasp* . . . are recruited during the on-line construction of metaphorical meanings" (273). However, it seems unlikely that physical meanings are recruited for interpretation as Gibbs maintains. Again, research reported by Steen suggests that this is not likely to be the case. Neurological analyses discussed by Lisa Aziz-Zadeh and Damasio also indicate that this is improbable.

At the same time, it is almost certainly true that physical meanings of "grasp" are at least briefly activated even with "grasp what he was saying."

(This would seem to account for some of the findings noted by Gibbs.) That activation probably does have some effects. For example, in some cases the activation may have emotional or associative consequences. Once some item is activated, we would expect it to have some consequences, for simple connectionist reasons. However, it does not appear likely that the effects are a function of modeling. There might be small biases introduced into thought by the partial activation of some circuit associated with physical grasping. But it does not follow that our thought about the target is organized and oriented by a physical grasp model. In addition, the activation here is not a function of metaphorical connections anyway. It has been well established for a long time that our minds activate various meanings of a word, including those that are contextually inappropriate (see Ingram 211 and citations). Thus, if I say "I rowed my boat to the bank," the word "bank" will briefly activate both "side of a river" and "place to keep money." This does not mean that "place to keep money" is in any sense a model for our understanding of the side of a river.

It seems that many effects attributed to body-based modeling are better understood as the result of bodily cognition. For example, research by Aziz-Zadeh, Stephen Wilson, Giacomo Rizzolatti, and Marco Iacoboni found "a clear congruence . . . between effector-specific activations of visually presented actions and of actions described by literal phrases" (1818). Findings of this sort are sometimes taken to support body-based modeling. But they seem much more in line with bodily cognition, as these researchers themselves indicate in their discussion of mirror neurons. Moreover, it is crucial that the congruence here was for *literal* phrases. In this research, "evidence for congruent somatotopic organization of semantic representations for metaphorical sentences in either hemisphere was not found" (38).

FLA: While I tend to keep at arm's length the embodied cognition approach, it is an undeniable fact that we grow our biology (brain, body, and all) within the social. It follows that with this we grow our capacity for creativity as socio-biologically grown organisms.

PCH: In connection with these senses of embodied cognition, the passage from Faulkner seems to evidence considerable bodily cognition, but its body-based modeling is much more limited. There are, of course, anthropomorphic characterizations of the bell and even the silence. But even that seems to have more to do with the misattribution of bodily experience than with drawing on models of bodily orientation. Thus the "tranquil" and "unhurried" qualities of the scene seem to reflect the sense that Quentin has when imagining himself at the bottom of the sea in the "cool eternal dark." Similarly, the hate that he attributes to Benjy is almost necessarily derived from mirroring (since Benjy has no means of verbally communicating hate). Finally, the "secret surges" are almost certainly his own feelings attributed to the salient object of his imagination—his sister and her lover.

FLA: A truly scientific theory of narrative fiction cannot be the sum of different theories or different hypotheses originating in all series of sciences or philosophies of mind or ideological constructions. A true science of narrative fiction cannot be the eclectic sum of debris taken from various disciplinary dustbins.

I think we should go back to the basics: what is our problem, and how do we solve it? The Russian formalists got it right: our task is to delimit the territory and develop the tools that will allow us to explore this territory. In delimiting the territory they made the important discovery of story and discourse. In my view, our task would be to rise to the challenges they posed.

PCH: Well, that nicely leads to the topic of our next conversation.

CHAPTER 3

On Matters of Narrative Fiction

FREDERICK LUIS ALDAMA: I have noticed that perhaps today with the use (and application) of cognitive neuroscience we have yet another approach to literature as *document*. Whereas in the past it might have been as historical or sociopolitical document, now it is as document for other purposes—or other sciences. As always, whether it is a New Historicist approach, a cultural studies approach, or even a strictly literary interpretation approach, there appears to be a confusion yet again of the fictional with operations that govern reality—the mind/brain or external world.

I think keeping the author-text-reader triadic model in the forefront of our work is absolutely necessary for the scientific study of narrative fiction. Otherwise, we begin to talk about characters as real human beings or themes, events, and so on, as if they are real—and not the creation of the author in the constituting or building of a blueprint with very specific aesthetic goals in mind. That is, *authored* products that cohere as a blueprint and that have a unified effect on the reader.

PATRICK COLM HOGAN: As you noted in the preceding discussion, we seem to disagree on the degree to which there is a gulf between fiction making and ordinary cognitive processes that engage the real world. You seem to wish to make a very sharp division there, somewhat insulating fiction as a strategic complex of aesthetic effects. In contrast, I see fiction as the result of ordinary processes of simulation—a connection initially made by Keith Oatley ("Why"). In other words, from my perspective, fiction making is exactly the same sort of thing as the activity of imagining what will happen if a person asks his or her boss for a raise. The same sorts of constraints relating to theory-of-mind presuppositions and the like are in effect.

We might think of the issue in the following terms. Fictions usually have both emotional and thematic purposes. The emotional purposes have to do with producing certain trajectories of affective response in readers. Note that the readers are in the real world, and the emotions they experience have evolved to deal with the real world. Thus it would be strange if they responded emotionally to, say, depictions of attachment that have nothing to do with the real world. For example, the opening of *À la recherche du temps perdu* has Marcel recalling the excruciating anxiety he felt when separated from his mother at bedtime. There are many affecting elements of this recollection. One of the most touching concerns his "only consolation," namely his mother coming up to kiss him goodnight. But "this goodnight lasted such a little time, she redescended so quickly, that the moment [he] heard her coming up" with "the light sound of her garden dress in blue muslin," his sorrow would surge again (21, my translation). Note particularly how Proust's phrasing suggests both a type of event repeated nightly in his childhood and also the particularity of a single memory, marked by the blue muslin garden dress. You are right about the separation of fact and fiction in that we should not assume that this is a detailed record of Proust's own experience. But would we really expect readers to be moved by this depiction (as I am) if it bore no relation to real attachment vulnerabilities, to common real-life patterns of anxiety, security, attentional orientation, and other matters?

The same point applies even more clearly to theme. A theme is precisely a means of connecting the fictional world with the real world. It is not clear that Proust is making any sort of thematic statement with Marcel's reminiscences about insomnia. But, if we turn to Tagore or to Harriet Beecher Stowe, we find a similar focus on attachment vulnerability with clear thematic implications. Take Tagore's story "Exercise-Book." In that story, a young girl is married and sent to her in-laws. She has one source of consolation in her loneliness—the possibility of developing her thoughts and creative impulses in a notebook. These thoughts and impulses prominently involve reflections on attachment loss, including reflections addressed to others. In a particularly affecting moment, she writes in her notebook a sentence that surely reflects the feelings of many young girls in her position. At home, she used to irritate her older brother. Now, separated from him when she is still a child, she writes, "Dādā [older brother], I beg you, take me home again just once—I promise not to annoy you" (143). This thematic commentary on child marriage is possible precisely because of the continuity between the fictional world and the real world.

This is closely related to the points I was making before about interior monologue versus what I referred to as "interior dialogue." When we simulate situations in ordinary life, we may simulate them with or without speech. If I simulate driving via a different route to work, I do so more or less visually. I do not formulate the simulation to myself in words. When I simulate speech, I most often do so in some context where I would be speaking—thus in relation to some simulated addressee. For instance, after I received some comments from you on our discussions, I went around the house doing ordinary things but responding to you in my mind. My "interior monologue" was not directed to me; it was directed toward a simulated Frederick. Indeed, it was much the same as what I am doing right now, since I am typing out a response to a comment you e-mailed to me. Note that the preceding sentence shows a second sort of simulation—my simulation of a reader. After all, you, Frederick, know that we are conducting our conversation in writing. I provided the information for the simulated "overhearer" of our conversation, a "side participant" as Richard Gerrig and Deborah Prentice would say. (As this indicates, "interior dialogue" as I understand it is different from the multiperson imagination discussed by Thomas Scheff in relation to Virginia Woolf.)

In this way, I at least to some extent accept the view that a literary work can serve as a document for the study of, say, emotion. Of course, we cannot merely accept a novel or poem uncritically and assume that its status and significance as data are transparent. But that is true for experimental research as well. Indeed, one of the great problems with empirical research in literature and related areas is that researchers start out with a hypothesis and simply interpret the data in a confirmatory manner, ignoring alternative hypotheses that the research may support equally well. The initial hypothesis guides the interpretation of the data and makes it appear that the meaning of the results is self-evident.

Indeed, I would like to briefly defend the idea that literary texts are documents for historical and cultural studies as well. This idea derives in part from critical discourse analysis and related developments, such as New Historicism. As you know, I feel that the particular approaches of these schools are often misguided. But there is a fundamental insight there that should not be lost. Our complexes of semantic associations are formed in part by social interactions, including mass media. These complexes partially guide our inferences and our affective responses. This gives rise to certain patterns across individual ideas and attitudes. Put differently, all our neural networks are to some extent unique. However, long-term social dynamics as well as short-term conditions will tend to partially align those networks. That alignment manifests itself in what Foucault and others refer to as *discourses* or, in another terminology, *problematics*. Thus there are patterns to the way that European Americans think about African Americans. The patterns are not simply governed by the facts (e.g., that, on average, African Americans tend to have darker skin than European Americans). These alignments are commonly intensified by institutions that, as Foucault puts it, govern who "may enter into discourse on a specific subject." Institutions permit such participation because the persons in question have "satisfied certain conditions" (224). Thus such social processes as publication selection tend to enhance spontaneous convergence of ideas and attitudes. The result of all this is that public speech and writing are very likely to manifest patterns of uniformity in the discourse of a particular time and place. For this reason, it can be very productive, indeed very important, to read literary works as cultural documents.

That said, I should add immediately that I fully endorse caution regarding the "documentary tendency," as we might call it. Despite their opposition to the idea of a consistent world picture (see, for example, Veeser xii–xiii), New Historicists and other writers in cultural studies often tend to assume that discourses are more uniform and more enduring than they in fact are—otherwise, they would not be so ready to draw connections between small details of Shakespeare's plays and small details found elsewhere in Renaissance texts (e.g., regarding some colonial action or some medical idea). They also tacitly assume that dominant discourses are more motivationally consequential. There are usually many discourses at any given time—and there are many partially inconsistent ideas and attitudes held by any individual and manifest in any given text. To some extent, we choose our discourses to fit our motivations rather than the reverse.

DEFINITIONS

FLA: I appreciate this further elucidation, Patrick. The documentary tendency in the analysis of fiction is so much the air we breathe in English department hallways that it is difficult to see it for what it is. To forestall any potential slips, perhaps a solid definition of narrative is in order. H. Porter Abbott's definition is simple, capacious, and sufficiently precise for our purpose here. Abbott says that "narrative is the representation of an event or a series of events" (13). And he explains that "the difference between events and their representation is the difference between story (the event or sequence of events) and narrative discourse (how the story is conveyed)" (15).

I would place at the center of narrative fiction what I call the "generative operator of discourse" that gives shape to the story. I prefer this definition. The generative operator as applied to story is also formal. The advantage of my definition is that from the start in the definition itself I am positing a very powerfully generative mechanism: discourse as an operator that shapes and transforms the story.

Authors select their narrative devices, such as focalization or free indirect discourse, and so forth, for specific purposes within the narrative and more largely to satisfy the aesthetic aims that the author has set him- or herself. What is interesting is that all the devices analyzed by Gérard Genette, Gerald Prince, Dorrit Cohn, and Seymour Chatman, among others, are inscribed within the very general category of discourse; they would all agree that these procedures belong to the domain of discourse. What I add with my definition of narrative fiction is the understanding that the properties of this very large domain we call discourse are *generative properties*. Discourse generates all those fictional signposts (free indirect discourse, discordant narration, psychonarration, and the like) that Cohn so cogently identifies.

That is, whatever we situate within the domain of discourse necessarily belongs to the domain of narrative fiction—and *only to that domain*. We do not find phenomena of focalization or free indirect discourse in the essays published by *Scientific American* or in handbooks on zoology, biology, and so on. Even if there is a story told, if that story is not given shape according to *aesthetic goals*—if the story is not submitted to this generative operator of discourse—that story *is not fictional*. And this is true even if that story is a lie. Even if something like Truman Capote's *In Cold Blood* (1966) is marketed as a true account, it remains a novel. It is as if we were saying that *Madame Bovary* (1856) is equivalent to a true-crime story just because it is based on a *fait divers* that really happened. Nobody would say that, even though some do suggest as much of *In Cold Blood* because Capote's declarations to the press made it murky. Capote isn't alone here, of course. I think readily of Gabriel García Márquez, who also made a mess of it when he said that magical realism *is the everyday realism* experienced by people in Columbia.

The moment you tell a true-life story using fictional means, the story ceases being a simple objective story about what actually happened out there. It becomes fictionalized because the use of fictional means automatically transforms, modifies, the true-life story.

Let me explain another way: the story can be a series of events that *actually happened.* The *fait divers* that Flaubert read actually happened. From the point of view of narrative theory, whether the story is true or not is immaterial. The author's starting point can be a *fait divers* or a story written by Edgar Allan Poe. The fictionality or nonfictionality of the prime matter is irrelevant. What automatically fictionalizes the story—whatever its origin, fictional or real, reality or fiction—is the application to that story of the operational generative device of discourse.

No matter Capote's declarations to the media of the factualness of *In Cold Blood,* the moment Capote uses discourse devices in the telling of the story, he automatically fictionalizes the story. The same with Flaubert. The moment he applies the discourse operator to the *fait divers* in the making of *Madame Bovary* it becomes fiction. Concerning the recent phenomena of docudramas, whether they like it or not, docudramas *are fictions based on real-life events.* The second the discourse operator is present, it is fiction.

PCH: I think we probably disagree on this. First, just to clarify what we are speaking about, we need to draw a further distinction. Discourse comprises narration and emplotment. Emplotment concerns the order, duration, and construal of story events. For example, suppose I tell you how I was detained by security at the airport after a scan found "an anomaly in the groin area" (as the guard delicately put it). I may begin by saying "Fortunately, they released me in time to catch my flight," or I may delay conveying that information until the end. I am guessing you would not say that any manipulation of emplotment—thus any change from chronological order—constitutes fictionalization.

So, it seems, then, that you are speaking of narration only. Now, we need to consider necessary versus sufficient conditions. Suppose you and

I are talking about what would have happened if Lalita and I had been hired at Ohio State. I might say that we would have bought a house near yours so that we could come over and improve our volleyball proficiency. That would be told in my ordinary narrational voice, but it would still be a fiction. So I think you are not saying that a shift in narrator is a necessary condition for something being fiction. Of course, you might say that the very fact that I am speaking counterfactually makes the speaker (the volleyball-playing "I") into someone other than the real me. But that makes the change in narrator dependent on the fictional status of the story. Thus it would still not mean that the narrative voice was the controlling factor in that case.

So, this leaves the possibility that manipulation of narrative voice is a sufficient but not necessary condition for fiction. My view is almost the precise opposite of this. My inclination is to say that all communication involves something like the creation of a narrative voice. The main difference between fiction and factual speech or (especially) writing is that we tend to adopt a fairly uniform voice in the latter but have much more leeway in the former. The uniform voice is largely a function of discourse constraints, including institutional constraints. Thus, suppose I actually speak a decidedly Scottish form of English. Even if that is my spontaneous way of speaking, I still will not adopt it in an academic article. In contrast, when writing a story, I might use my spontaneous speech patterns. In this case, then, the manipulation of a narrative voice is much clearer in the academic article.

THE DOMAIN OF NARRATIVE (I): CONTEXTUALISM

FLA: There have been various attempts to develop what is called a contextualist or postclassical narratology: feminist, ethnic, cognitive historicist, and so on. When talking about context we are actually talking about adding to a strict science of narrative fiction whatever perspective the scholar considers necessary.

So what happens is that scholars assume that the science of narrative fiction does not have *its own exclusive and proper domain,* its own proper exclusive field of study. I wonder if all this contextualism is a winding back of the clock to before the formalists, to a point in time when it was said that the study of narrative fiction is not a science, not a clearly delimited domain of study. Rather, it includes everything and anything.

PCH: This is related to the issue of so-called political correctness. As you know, I share your impatience with many aspects of this, and we go into this topic more in chapter 5. There seem to be three issues here. One concerns the validity of the contextualist approaches. Another concerns the validity of the scientific approach. The third concerns the institutional and professional structures that underlie the relations between the two—this is what bears on "political correctness."

I want to focus on the final point. I think we would both agree with Gerald Graff that the issue here is not so much politics as pseudopolitics or a sort of political rhetoric that usually has little or no bearing on the actual conditions of people's lives. It is, rather, a form of bad faith, in roughly the Sartrean sense (see Sartre 87ff.).

To put the matter crudely, the structure of the profession requires that we argue in favor of intellectual positions that afford us greater publication opportunities. So, you and I do not only have intellectual commitments to cognitive study. We also have class commitments, commitments based on our position in political economy. Specifically, as cognitivism becomes more widely accepted, we are more likely to have our writings accepted for publication and thus receive merit raises in our institutions, as well as other benefits. Indeed, the structure of the system is such that one almost necessarily has to adopt some intellectual identity category in order to advance oneself. There are many ways in which I am as close to Foucault as I am to Damasio. I certainly have agreements and disagreements with both. Indeed, I am no less dissatisfied with much cognitive literary criticism than I am with much cultural study. But, as a matter of institutional fact, my "allies" are in cognitive study. No Foucaultian will invite me to be a plenary speaker at a conference; few would agree that a book of mine should even be published.

Everyone is in this position (though, of course, the categories vary). But it simply will not do if we all go around speaking of alliances. Rather, discourse practices constrain justifications for our action and argument. Specifically, there must be some degree of nonegoistic merit claimed by our arguments. There seem to be two commonly accepted forms of merit. One is validity. The other is productivity. Productivity may be further subdivided into prudential and normative (either ethical or political). So, there are basically three forms of argument available to us. The first is that our claims are true (while those of our opponents are not). The second and third are that our practices have benefits. The benefits may be a matter of practical success, or they may be a matter of normative (ethical or political) improvement.

You and I are in complete agreement, I believe, that validity should be the fundamental argument. If nothing else, it should at least qualify claims of improvement. In other words, suppose we are going to justify everyone adopting deconstruction because it undermines patriarchy. Then we should at least have a basic concern with whether or not deconstructive claims really do undermine patriarchy.

On the other hand, I think we would agree also that "truth" is (as Foucauldians and others have stressed) widely invoked as a way of justifying political oppression or other sorts of prudential or normative harm. Thus it is important to be aware of the political implications of claims, the degree to which arguments and analyses conform to standard discursive structures or dominant ideologies, and so on. In other words, we do wish to privilege arguments based on validity. But we would not want to simply assume the good faith of writers who invoke truth—or the bad faith of writers who invoke norms.

More concretely, since the 1960s, literary studies has developed in such a way as to require that faculty publish more. This has intensified the institutional pressure to strategically advance one's alliances. In keeping with this, there has been an increased political awareness that is in part both sincere and important, the result of the great popular movements of the 1960s. At the same time, this political awareness gave particular salience to political norms as legitimation principles. Thus there was, for many years, a strong tendency for literary critics and theorists to support their schools of thought by claims about political benefits. In some cases, these claims were at least partially true—as in the expansion of the canon to include women and nonwhite authors and the expansion of the faculty to include women, nonwhites, gays, and lesbians. But there was also a certain amount of bad faith in all this—not because the individuals involved were sinister but because the institutional structure leads to discursive constraints that foster bad faith. Thus there have been legitimate complaints about the valuing of normative justifications over justifications based in validity (e.g., in valuing deconstruction on the grounds that it is antipatriarchal rather than because it rests on a plausible semantic theory). At the same time, the opponents of such normative arguments have their own institutional positions, self-interests, class interests, and thus bad faith. Indeed, the very idea of "political correctness" is a way of dismissing both the valid and invalid aspects of the normative (or contextual) approach.

So, what does all this point to? It seems to point to a situation in which what we really want to do is reduce discursive constraints as much as possible. The valuable parts of the contextual approach—the parts that seek to open the field to greater diversity—try to do just that. But the problematic parts do just the opposite. They try to stifle opposition, by what is often little more than name-calling (even if the names are fancy, such as "logophallocentrism").

There is another aspect to this problem that is more psychological than political. Humans seem to have a broad tendency to feel that their own behavior or preference is legitimate only if everyone does exactly the same thing. Thus we seem to want everyone to do the same sort of literary theory, to share the same literary canon, or—to take a particular irritation of mine—to teach in the same way. (Evidently, wide-ranging class discussion is what we are all supposed to do—with, of course, no consideration of what methods seem to produce greater learning.) Thus it appears that we need social institutions that foster different approaches—here, scientific as well as contextual approaches to narrative. At the same time, we need to have a personal sense that our own approach need not be shared by everyone.

THE DOMAIN OF NARRATIVE (II): STYLE

FLA: Nicely put, Patrick. I'm certainly not going to stand in the way of anybody's source of pleasure (contextualist or otherwise). There are many branches to harvest fruit from here.

I happen to take my pleasure in a research program focused on getting at foundations—the roots, if you will. While we go at this differently, in our works we seek to push more and more at determining the contours of a research program that will allow for a foundational understanding of narrative fiction.

With this in mind let me turn to a concept I bring up in chapter 1. Here I mention the narrative as blueprint. Let me go into a little more detail here about what I mean. All novels, short stories, films, comic books, and the like are made according to recipes, algorithms, or blueprints made up of such ingredients. The concept of the blueprint aims not only to capture all the ingredients that make up a given narrative fiction—from technical devices and structures the author employs to plots, events, and character dialogue and action—but also to convey the sense that all these elements that make up the respective narrative fiction are put there by the creator (or creators, in the case of films and comic books) who seeks to produce a work in which all the elements cohere and achieve a unity of effect in relation to the audience.

Creators of narrative fictions discipline their emotion and cognitive systems in the skillful creating of the blueprint (imagination plus the purposeful use of technique) so as to engage an ideal audience that the creator assumes shares basic sensory, emotive, and cognitive faculties.

Here and elsewhere I use the term "will to style" as a shorthand to identify the degree of presence of willfulness in the creator's use of technique and imagination and his or her responsibility to subject matter in the crafting of the blueprint. Of course, this "will to style" varies greatly from product to product.

PCH: Style is, of course, a key issue in verbal art, Frederick. Here, I would distinguish verbal art from fiction. The key feature of verbal art does not seem to me a matter of truth claims. In a very traditional way, I would tie the definition of fiction to truth claims, not narration. Specifically, I would say that a work is fictional to the degree that it is presented as not claiming its verbal or visual depictions to be true. Thus a historical fiction would be fictional in certain respects and nonfictional in other respects. (There is a separate issue of whether the nonfictional aspects are accurate or not.) By this definition, a wide range of narratives are to some extent fictional and to some extent nonfictional. A more definitive division may be made in the following way. The default value of a work of nonfiction is truth. In other words, any given statement in a work of nonfiction should be assumed to have claims to truth unless there is clear indication to the contrary. In fiction, this default does not hold. (Note that this does not mean that there is a presumption of falsity. The case of historical fiction shows this clearly.)

In contrast, I would say that verbal art is a form of speech or writing in which there is an increase in the possible relevance of all aspects of the utterance to emotional or thematic purposes. This is where narration and emplotment enter, as well as style. Aspects of story order and construal, aspects of projected narrative voice, aspects of phrasing (including even sound patterns)—these enter into ordinary communication but only in limited and peripheral ways. For example, as to style, we might try to wittily make a rhyme in telling something; conversely, in writing a scholarly essay, we might try to avoid a potentially comic sequence of alliterating words. But we typically do not spend much time on these aspects of writing or speech if we are not engaged in producing verbal art.

Verbal art, then, may be defined by this intensified concentration of relevance. Note that this means that there will be a gradient of verbal art just as there is a gradient of fictionality. Moreover, the two will not be unrelated. We are clearly more free to manipulate plot, narrative voice, even style, if we are not constrained by a default presumption of factuality. For example, people generally speak rather unpoetically. If I am constrained to report people's actual dialogue, I am not free to manipulate the style of that dialogue toward aesthetic ends. On the other hand, it is clear that history or biography can be written with more or less attention to style.

FLA: I like your formulation of verbal art as an "intensified concentration of relevance," Patrick. I agree that this can be shaped by aesthetic or nonaesthetic goals. In many ways, it is our capacity to intensify any elements of our language (narrative fiction, poetry, or that article in *Scientific American*) that ultimately led the Russian formalists into dead ends. There was nothing special about the language in poetry that would mark it as definitively different from that of everyday usage.

I know we cover the topic of linguistic approaches extensively in chapter 2, but it is worth mentioning the following. With the exception of some forms of poetry, in almost all forms of literature, language is actually immaterial. Just like we say story as such is immaterial—we do not care if it is based on a true story or one by Poe; what matters is that it is a matter to be shaped and transformed by discourse. David Malouf's *Ransom* (2009) and Zachery Mason's *The Lost Books of the Odyssey* (2007) both pick up stories in the *Iliad* and spin new stories. Ninety percent of what Borges wrote was based on this principle—he takes from anywhere, including his own expository essays, and applies this generative operator of discourse and creates something completely new.

Dostoyevsky wrote in Russian. I cannot read him in the language in which he wrote. All that I know about his novels, all that I have enjoyed, all that has given me an aesthetic satisfaction in Dostoyevsky's work, has no relation to the Russian language. There are many authors in Latin America and Spain who know not a single word in English. Yet they experience the full impact of writers like Dos Passos, Hemingway, and Faulkner. García Márquez, for example, does not know much English. He has read all these authors in Spanish. He says that if it weren't for this handful of authors, he would not have been able to shape his fictions the way he has.

Besides the enormous confusion of what we mean by linguistics and its use in the analysis of narrative fiction and then the enormous confusion of what we mean by the *level of generality* or *level of abstraction* at which we enter into linguistics as a science and confusion between theoretical and applied—as we discuss in chapter 2—there is the confusion that directly concerns the place of language *as such*, first, within the empirical experience of narrative fiction, and second, and most importantly, *within* a scientific theory of narrative fiction.

I am proposing that the two pillars of narrative fiction are story and discourse. The language in which the narrative fiction is conveyed is a secondary consideration. From this point of view, it seems to be a misdirected effort to try to explain at a foundational level narrative fiction in terms of language, language use, and language study.

What is really important is all that is inside the toolbox of discourse—Cohn's signposts of fictionality, Prince's narrator–narratee constructions, the different instruments of analysis identified by Genette such as frequency, mode, and duration.

PCH: There are two separate issues here. The first concerns the bearing of linguistics on aspects of a literary work that are not directly linguistic. The other concerns the importance of language generally in fiction.

As to the former, the relevance of linguistic theory to narrative study depends on the generalizability of features of the theory outside of language. The point of Structuralist generalization is not that language proper is everything. Rather, the point is that social systems have the same structure as language. Thus, to take standard examples, we might find that the meaning of myths is a function of binary oppositions or that the production of dreams is organized by axes of substitution and concatenation. The problem with Structuralism, I think, was not that it gives language per se too much importance. In fact, Structuralist analyses often ignore language. Rather, the problem is that it both overgeneralizes the applicability of language structure to other fields and begins with a wholly inadequate account of language structure.

In fact, I find the same problem with Genette. In *Affective Narratology*, I consider time differently from Genette. This is because I do not begin by assuming that the temporal organization of narrative is fundamentally a function of linguistic tense or aspect. Of course, in Genette's case, the basic distinctions of order and duration are valid and important. (I am less sure about frequency.) The problem here is primarily that they do not exhaust important features of narrative time.

As I suggest in chapter 2, I do believe that some aspects of linguistic theory have promise for literary study—if we are careful about what is likely to generalize. Thus some features of linguistic processing isolated in psycholinguistics are likely to be common to other forms of cognitive processing. For instance, if our language processes are well represented by principles with parameters, it seems likely that principles with parameters will turn up elsewhere. Similarly, some group dynamical features isolated by sociolinguistics are likely to be shared by other forms of social interaction.

As to language and narrative, here too we need to make some distinctions. The first is a distinction between the literary work and its narrative. It is true that the majority of what we consider a work's narrative will be preserved across changes in language or even medium (e.g., from prose to film). But that is a sort of trivial point. It merely says something about the way we use the word "narrative." We use it to refer primarily to story and secondarily to discourse, particularly emplotment. We tend to count language as bearing on narrative only insofar as it changes key features of the story. In short, the relative irrelevance of language to "narrative" is mostly definitional. (Actually, I myself think that the language makes a big difference to the story. But that does not seem to be the standard operational view.)

The more consequential question, then, concerns the importance of language to literary works. But once it is phrased like this, the banality of the basic answer becomes obvious. The basic answer is, of course—it depends on the work. Sometimes people try to make generalizations, and those can be approximately valid. For example, it is generally the case that the language is more important to poetry than to prose fiction. But even then one has to make case-by-case decisions.

You bring up translation, and that points to another set of issues. Here, we need to consider what is important about the language of the source work and the degree to which those features may be transferred to the language of the target work (that is, the translation). It is a commonplace to remark that one nontransferable feature is sound. That is already a problem. The problem is intensified when sound is brought into relation with other features. For example, we may be able to imitate the sound pattern of a passage or convey the sense. But we typically cannot do both together. Moreover, the problem goes well beyond sound. It turns up with the complexes of semantic associations that happen to have accrued to a particular word or phrase, what the Sanskrit aestheticians called *dhvani*, or suggestion. This becomes particularly important when that cloud of associations is connected with a particular complex of feelings, or a *rasa*. For example, in Premchand's *Godān*, there is a very moving scene in which the main characters, Hori and Dhaniya, are faced with the woman whom their son has impregnated. She has come to their house, and Hori has said he will throw her out. However, she appeals to him, pathetically throwing herself at his feet. In Hindi there are three forms of "you"—formal, informal, and intimate. The girl addresses Hori with the intimate form, thereby conveying the sense that she is like his daughter. When Hori, moved by her entreaty, calls her "beti" and says that this is "tera ghar" (152), this conveys a tenderness that is missing in the English equivalents—"daughter" (as a form of address) and "your house" (using the intimate "you"). We probably would not say that these points in the language are crucial to the narrative—simply because of the way we use the word "narrative." However, they are, to my mind, crucial to the work.

FLA: Of course, language is the prime matter, say, of narrative fiction; in film it would be visual images and auditory sounds, and in comics it would be a combination of the verbal and the visual. We can choose to pick at the branches of the infinite number of iterations of language in narrative fiction, or we can move toward that work in biolinguistics that offers a high level of generality that can potentially shed light on an equally high level of generality in a formulation of narrative fiction making and consuming. The generative operator of discourse offers such a level of generality. It is here that we can better understand the work of Cohn aimed at identifying the signposts of fictionality—all these signposts are the devices used by what I am calling the generative operator of discourse. They are indications of the fact that no matter the origin or nature of the story, the moment it is fictionalized, it makes use of devices such as free indirect discourse or psychonarration.

Discourse is the generative operator that generates an infinite number of stories when applied to the story or historical structure. This is so because this operator is constantly giving shape and acting on the story element and giving it a potentially infinite number of shapes.

The story of an adulterous woman is the story of Emma Bovary and also of Anna Karenina—yet they are totally different novels. Why? Because of the shaping activity of the artists—their decision to situate their novels within a specific time and place and to tell their story using different sorts of human interactions. Flaubert's choice, for example, to have Emma doing the proverbial beast with two backs in the carriage with Léon all while enveloping her within a particular social fabric makes this a perfectly individualized story that is different from all others, including *Anna Karenina* (1873–77).

94 • CHAPTER 3

PCH: These are certainly significant—and complicated—issues. You are undoubtedly right that discourse manipulation is often an important factor, though your examples may not be perfectly well chosen to illustrate that. It may make more sense to consider, say, *Jane Eyre* (1847) and *Wide Sargasso Sea* (1966). One difference between these two works is a change in narrator and thus an element of discourse. On the other hand, as I have already said, I am not sure that discourse has the "operator" function you see it as having. Indeed, I see the definition of a narrator itself as a simulative operation. The point is particularly clear in the cases of these novels where we are dealing with character narrators.

CHARACTER

FLA: "Why do we care about literary characters?" Blakey Vermeule has asked recently, as have many others before her. Fictional characters are simply speaking constructs. So in the domain of character analysis and story analysis, everything is man-made, so to speak; it is artificial. They are not domains of discovery. They are domains of artifice. We only care about characters, when we do care, when the author has succeeded in creating the characters in sufficiently clear-cut and at the same time schematic and complex ways.

I just watched Robert Rodriguez's *Roadracers* (1994). The protagonist Dude Dudley (David Arquette) makes certain extreme, menacing facial expressions at the end of the film—and not at the beginning. Rodriguez's choice of lens exaggerates the expressions further, giving them a kind of cartoon effect. The decisions here as well as those relating to how the characters interact and the choice of actors and costumes (identifiable good guy and bad guy) all give the story its shape. Rodriguez chooses to give his characters schematic shape by painting them with huge brushstrokes and endowing them with cut-and-dried personality traits. He omits all of the complexity of human beings. All we know about Dude is that he wants to get out of town, and that's all we need to know for the film to work. But we can say the same of more complex characters, too. What do I know about Raskolnikov? I know he has a desire to kill the pawnbroker and use the money he plans on stealing from her to pay his rent and finish his studies. I know that he feels affection and falls in love with the fallen Sonia. But what do I really know of all the motivations, psychological traits of Raskolnikov—or an Anna Karenina or an Emma

Bovary? In actuality, I know very little about their psychological makeup as compared to any human being in our real everyday life.

This is what art is. The awe-producing activity of the verbal artist is precisely that he or she is able to select those psychological traits that are most relevant to the development of the story and that will reverberate in our memory long after we turn the last page.

Everyday life functioning of the brain is messy, complicated, and mixed. It is shapeless in every way, except in its functions governed by laws of nature (physics, biology, etc.). Art by definition *is shape*. The main activity of the artist, therefore, is the giving of shape to his or her chosen object. It is a shape-giving activity triggered and guided by aesthetic goals. All human work is teleological; it is guided by aims or goals. The carpenter's goals are to manufacture tables and chairs. The artist's goals are to create aesthetic artifacts. The artist aims at creating certain emotional and cognitive, affective and rational responses or reactions to the products of her work. Those reactions are generally described as being aesthetic.

Let us take the example of *Madame Bovary*. Flaubert reads in the section of the newspaper concerning crimes ("la section fait divers") about an adulterous woman who kills herself. That's the story: a married woman has sex with men other than her husband, gets herself into debt, and commits suicide. If our fiction capacity were reduced to the faculty of inventing stories, then the novel *Madame Bovary*, no matter how much the author stretched sentences, would tell a very simple story, with very little complexity. What turns this story into a work of fiction and a work of art is the discourse, the injection inside the story of shape. In this case, the story triggers a whole series of faculties in Flaubert's mind, and he chooses to talk about the psychology of Charles, Léon, Rodolphe, Monsieur Homais, Emma—all in *a very selective way:* he ascribes to Léon such-and-such definitional psychological traits and the different traits to Rodolphe, allowing us to distinguish one from the other. When each of their names appears it triggers in our mind a cluster of psychological traits associated with each of the lovers. Once the traits are selected respective to the characters, the time and place of the story, and the central theme, then Flaubert decides how this will be conveyed through language. All the choices give shape to the fiction.

Art is by definition giving shape to something. Except for the sunset and all other iterations of the sublime, all aesthetic objects are necessarily shapes. There is no shapeless art. (In this sense, I am rather skeptical about conceptual art—to me it is not art.) Art is the deliberate, painstaking, detailed conferring of shape to matter—to a material reality.

The thematics, the scenery, passage of time—all this is shape. And all this shaping is built into the story by the generative operator of discourse. What attracts us to characters and what attracts us to narrative fiction is not the story per se. It might be true that the number of stories that can be told amounts to a handful or so, but because of the generative property of narrative fiction we have an infinite number of iterations.

The generative operator of discourse in this sense works in parallel to the generative capacity in language that turns the building blocks of language into a potentially infinite number of applications of language that are totally different, one from the other.

When we read a novel (even those that are poorly constructed), we know the character in ways that we could never know a person in real life. This is so, of course, because an author selects each and all of the psychological traits ascribed to the character. And when an author chooses to include an enumerable list of psychological traits they will always be a finite. It could be one or ten or one hundred traits, yet we still perceive them as separate because of their particular form, their particular shape, and their particular function within the story as a whole. With respect to a human being this is impossible. We never have the impression that we have a full 360-degree view of anybody's character or that we perceive a unified whole made out of a finite list of psychological traits and functions. Even the most boring, elementary-seeming person is a source of surprise. The only way a character may surprise a reader is if the author *decides* that this character will behave or react in what would seem a surprising, unforeseen way—and this for all eternity.

Real-life human psychology (unlimited) has nothing to do with narrative fiction psychology, where there is always a selection of psychological traits made according to aesthetic goals, traits that guide the application of a whole series of narrative devices.

So when scholars of literature talk about characters as if they were real-life human beings, they seem to be way off the mark. Dostoyevsky decides what psychological traits to assign to Sonia, the father, and so forth. Hemingway decides what is shown and not shown, what is underneath the water level of the iceberg. Authors decide how the reader is to fill gaps through the application of all their mental faculties—by inferring, calculating probabilities, using causal reasoning, plus many other capacities that also concern the emotion system. Authors decide what they will show and not show in the motivations and interactions among characters. In the hands of creators of narrative fiction, there are a seemingly limitless number of ways conferring such shape.

Whatever is in the story concerning characters and many other aspects, *the author has put it in there*. It is the product of the author's "will to style"—that will to achieve certain aesthetic goals through specific means. The author chooses everything inside the narrative fiction: the degree of complexity of mental operations and relationships of the characters and the number of elements from social institutions that interface with the political, sociological, economic, historical, and so on.

Narrative fiction is infinitely more shaped than reality, with its lack of aesthetic boundaries. Since I am not a character in a narrative fiction, that is, since I am a real-life person, I am strictly the product of an accident. Born to Luis Aldama and Charlotte Ann Robles-Castillo Aldama, I could have been born years later in 1984 and in Columbus, Ohio. The fact is, our lives are almost totally shapeless. The little shape they have originates in the fact that we are part of nature. We are socio-neurobiological and historical organisms. On this basis, we give our lives shape according to circumstances, by behaving in ways A, B, C in response to this accident, by reacting to situations A, B, C, where both the circumstances and the reactions to them are not predetermined or preordained.

All is trial and error in real everyday existence. All is predetermined in narrative fiction, including that which occurs by accident. For it is the artist who decides what to keep and what to delete in his work, even those items that appeared unplanned, unforeseen, unwilled in the text. Even in such cases, it is up to the writer to decide if those items will be shaped into the global and final version of the work. The author will decide where those fortuitous items will be placed, how specifically they will be shaped, and how they will fit organically in the whole.

This is common sense, but it is also in the evidence provided by authors themselves. There are so many written testimonials of authors working through how they shaped novels, expressing surprise that people could think otherwise.[1] If we take a quick look at Dostoyevsky's journals on the writing of his novels, we can see how the most dazzling scenes in *The Brothers Karamazov* (1880), *Demons* (1872), and *Crime and Punishment* were scenes that he worked through very carefully to achieve a unity of effect. The same can be seen with Flaubert or with Philip Roth. There is not an event, phrase, a word even, that is not the product of creators'

---

1. See Aldama's *Spilling the Beans in Chicanolandia* and the interviews in Aldama's *Your Brain on Latino Comics*.

shape-giving activity, an activity guided and motored by aesthetic goals and aspirations.

There is no mystery here. There is no ontological conflation here.

PCH: This is interesting. We seem to be poles apart here. You feel that we know less about characters than about real people. My view is that we know far more—though obviously there are exceptions in each direction. For example, you say that we do not know much about Raskolnikov except that he wants to murder an old woman and use the money. In fact, however, we know all sorts of things about him. We know how he goes about planning the murder, what philosophical issues arise for him, what his emotions and reflections are at the time of the murder and just after (e.g., "Suddenly he wanted to abandon everything"; then "He was even beginning to laugh at himself," but this is cut short by a sudden imagination "that the old woman was still alive" [84]), and so on. We have hundreds of pages that reliably tell us what he is thinking and feeling. We know vastly more about him than we know about any actual murderer. Moreover, we know it far more reliably. The same point holds for Emma Bovary.

The issue you are pointing to is, I think, somewhat different—and different in the cases of Rodriguez and Dostoevsky or Flaubert. Specifically, in real life, particularly when treating other people, we tend to be very reductive in our accounts of motivation. We recognize that we ourselves respond in complex ways to changing conditions. But we tend to seek simple, character-based explanations for other people's behavior (see Holland, Holyoak, Nisbett, and Thagard 222–24). The tendency is almost certainly enhanced with out-group members, since we tend to view out-groups "as relatively less complex, less variable, and less individuated" (Duckitt 81). Murderers would almost certainly define an out-group for most of us. When thinking about murderers, we want some straightforward explanation for their behavior. The usual explanation is that they want money. So we reduce everything that goes on with a murderer to some simple formula, most often a standardized explanation. (Part of a standard discourse on criminality, to return to Foucault.) We do this not because we know more about the murderer, but precisely because we know so much less—and we know it less reliably. Moreover, our information is less salient; thus we can dismiss complications more readily. The point of knowing more and knowing better is precisely that it makes simplifying and reductive explanations less plausible. That is why you are completely right to cite Raskolnikov's interest in the money. It is entirely clear that this is an inadequate explanation—particularly given what happens afterward. So the case is very well chosen. However, it is well chosen in such a way as to show that we know Raskolnikov too well to attribute such simple motives to him—not that we know him too little.

THE DOMAIN OF NARRATIVE (III): DISCOURSE, SELECTION, AND SEGMENTATION

FLA: I see your point, Patrick. I would put the emphasis on the author's or artist's willful selecting of details (more or less) to create a highly aestheticized effect of the mind as well as that of time and space. The moment time and space are fictionalized, they cease having the properties of time and space identified by physics. Take for instance Cortázar's short story "The Continuity of Parks," in which the "real" space of the character-as-reader collapses into the "fictional" space of the novel he's reading—a novel that concerns a jealous lover about to kill a man reading a novel. As the story comes to a close we realize that within the diegesis the character-as-reader and the lover of the novel being read suddenly share the same ontological space.

This interpenetration of fictional world with real world is a physical impossibility. Under any theory of physics, quantum mechanics or otherwise, there is no way that these two ontological levels can coincide: the level of a fictional character in a story interpenetrating the level of another fictional character that is presented or posited as being a real-life reader. When reading a novel or watching a movie, those characters do not obey the laws of nature; the laws of nature themselves become fictionalized.

Whether it is in a more economic and artificial way, as in a Rodriguez film, or in a more sophisticated way, as in Cortázar's story, or whether it is the use of natural locations, as in the film *Deliverance* (1972), where all takes place *out there in nature,* this way or in any other way, time and space do not follow the laws of physics; they follow the narrative conventions and the aesthetic goals fixed by the author (director, etc.). Imagine all the events taking place in a story that happen over days or years that have to take place during the screening time of a film. No matter how "realistic" the locations are, those locations are always fragments of a space. Space contracts or expands according to the director's needs and aesthetic goals. This could be a simple car chase that's depicted or the space depicted in *Deliverance*—it is not a continuous space; it is constantly cut (fragmented) again and again according to the needs and aesthetic goals of the creator.

There are many devices put to use to operate on space. In cinema we have editing, but we also have the way the camera focuses (close-up, medium, and long shots, etc.) on the characters and landscapes and the different lenses (wide angle, telescopic, etc.). Space is constantly manipulated by all sorts of narrative devices in all the media, including novels, short stories, comic books, and so on.

And, we can consider many fictions that emphasize the manipulation of time in otherwise ontologically impossible ways. Think of Alejo Carpentier's story "Viaje a la semilla" ("Journey Back to the Source," 1944) and Martin Amis's *Time's Arrow* (1991), where time moves in reverse. Whether space or time or both, physical reality has a completely different ontological existence in narrative fiction.

Even in the most so-called realist films or novels, the shape-conferring properties of discourse necessarily bestow on the fiction an artificial temporal and spatial order and dimension. So, necessarily, the moment the time and space ingredients of the story become part of the diegesis, they are no longer the *real* time and space, the time and space governed by the laws of physics. The moment time and space are subject to the operation of discourse, they necessarily become *fictional* time and space—in all instances, even the most realistic instances.

Authors shape time and space in an infinite number of ways. They can slow down and accelerate the story through a whole series of devices related to the "order in narrative." These devices concern centrally the dimension of time. It is the generative operator of discourse that allows for disjunction between the time of the story and the time of the narrative. This differs from physical time that we can and do segment but in regular ways. In narrative fiction we have time that is segmented and more complexly organized and that is potentially infinitely variable. Through a narrative an author can condense one phrase in ten years or expand twenty-four hours to fill out five hundred pages as in the case of *Ulysses*. A narrative could even spend the same amount of pages narrating the biography of a character before that character is born, as with Sterne's *Tristram Shandy* (1759). Then there is the organizing of the space of the story whereby the authors can use different devices, such as the choice of point of view, for instance, that is always necessarily situated somewhere.

PCH: We seem to be moving along somewhat parallel tracks here. You say, "It is all about discourse" and I say, "It is all about simulation"—or, at least, the particularization of any imagined trajectory is all about simulation. In addition, you keep returning to the fiction–nonfiction axis, which seems to me much less significant.

First, I should set aside one issue. I disagree with the view that second-person narratives are addresses to the real reader. In my view, they are references to a simulated reader within some level of the storyworld, with only very limited exceptions that are highly context-specific. For example, a lyric poem from Li Ch'ing-chao to her husband will incorporate the real "you" as addressee when it is sent to her husband but not otherwise. In contrast, Calvino's *Se una notte d'inverno un viaggiatore* ... is not addressing any reader. Indeed, the point should be obvious. The "you" in the novel is doing things that roughly correspond to what is being done by the real reader. But it is clear in most cases that this is only because reading a book tends to have some recurring features. The details differ greatly. Anyone who is at home alone has no "others" to address (as Calvino's text assumes); for anyone who is in a library, there is no issue of a television (which figures in Calvino); and so on. The point is even obtrusive when Calvino details the bookshop with its delightful book categories—wholly irrelevant for a bookshop but quite familiar to avid readers (such as "Books That, If You Had More Lives to Live, Certainly These Too You Would Willingly Read, But Unfortunately The Days You Have To Live Are Only What They Are" [5, my translation]). Thus, I would rather say that there is a feeling of paradox or impossibility produced by such works. But it is not because there is some sort of crossing into the space of the reader or sharing of ontological space.

On the other hand, I of course agree that a fictional storyworld need not follow the same spatial and temporal principles as the real world. But, again, this does not seem to me a matter of discourse. Discourse reordering or changes in duration are continuous with our fragmentary memories, our ordinary (nonliterary and nonfictional) storytelling, and other facets of ordinary life. (In this way, I hold fully to Monika Fludernik's general advocacy of a natural narratology, a narratology that sees practices of narrative verbal art in relation to those of ordinary life.) More significantly, small parametric changes are precisely what define processes of counterfactual simulation. Yes, it is true that we feel free to vary some aspects of the world in fiction that we do not ordinarily vary in everyday speech. But counterfactual simulation is precisely the process of isolating one or two parameters and saying, "If only q were not the case, then I certainly would have gotten that job" (or whatever). The same point holds for hypothetical simulation, as in "Well, I still have a chance of getting that job if p happens." That sort of simulation is precisely what is involved when an author imagines that, say, the murderer in a story is "the reader" (a simulated agent, not me or you).

FLA: I like that we are running on parallel tracks, Patrick. It's certainly helping sharpen my thoughts on how narrative fiction works. On this score, I would like to continue along this track of segmentivity. I have come across theoretical work that I believe is very fruitful. It is centered on the concept of segmentivity and in the characterization of poetry according to the principle of segmentivity. Brian McHale discusses this in "Beginning to Think about Narrative in Poetry," where he brings to the fore this potentially powerful characterization as found in Rachel Blau DuPlessis's 1996 published "Manifests." We have with DuPlessis's characterization and McHale's astute application of "segmentivity" a powerful new tool for understanding how segmentation works to give shape and meaning to poetry. DuPlessis argues that poetry as a genre "is just selected words arranged by segmentation on various scales" and that it "is the creation of meaningful sequence by the negotiation of gap (line break, stanza break, page space)" (51). Following these general characterizations of poetry as a genre, DuPlessis offers a more specific definition:

> Poetry is the kind of writing that is articulated in sequenced, gapped lines and whose meanings are created by occurring in bounded units precisely chosen, units operating in relation to chosen pause or silence. (51)

The poet's willful use of space design (segmentation) whereby he or she jumps from one line to the next (and from one stanza to another) can emphasize last and first words of each line in ways that generate a semantic significance and energetic charge. Such a will to segment can sometimes even lead the poet to slice words up for kinetic force.[2]

I would add that segmentation might also prove a powerful tool in understanding narrative fiction. I am thinking of not only the paragraph and chapter breaks in novels but also the segmentation that gives shape and meaning in comic book (panels) and film (editing techniques). I think it might be a productive heuristic that we can refine, eventually make airtight, and then apply to all variety of narrative fictions.

PCH: That makes sense. As cognitive scientists have noted, our minds organize the world in part by selecting information, breaking it into units for

2. See Aldama's *Formal Matters in Contemporary Latino Poetry*.

processing, and bringing those units into structural relations with one another. One key unit in literature is the poetic line, presumably guided by the structure of working memory.[3]

I am a little uncomfortable with the metaphor of a "tool." Tools make products, and that is, of course, what we do in producing interpretations. We craft commodities that we then put up on an academic market for sale. Something is a valuable tool insofar as it facilitates such productivity. But that is an unfortunate aspect of the political economy of our profession, as I am sure you would agree. In consequence, I would rather ask if something is a valuable part of an adequate description of the object being studied and if it contributes to a plausible explanation of the relevant phenomena.

FLA: Point taken. I do think "tool" is a useful shorthand for describing all those devices available to authors—and we're constantly discovering new tools for the toolbox—as generative devices of discourse that shape the story. Let us move on to a more foundational question—that of literature as part of culture making. Culture is a composite of the material and intellectual products and their interpretation. This aggregate or global definition of culture as being all material products (socks, food, buildings) of human activity plus all intellectual products (narratology, Bayesian probability, or, $E = mc^2$?, etc.) comes with an attribution of meaning, significance, or importance *and* interpretation. Culture is always material *and* constantly submitted to interpretations. And cultural products are distinguishable because of their capacity to generate meaning, meaning in the most general and nontechnical, as opposed to the mathematical and logical, sense of the word. Cultural products are also systematically assigned a significance—a degree of importance. If I am a religious zealot, then I will say that the most important values concern religion. There is a scale of values that changes among different people. And among different people, certain scales of values coincide—all those who are zealots can be put in one group, others who are not fanatical in another, and so on and so forth—and the parts of groups can overlap.

It is quite useful to have an approach that understands that culture is the unity of material, intellectual, and interpretive activities. Within this sense of culture we find also that worldviews are born and given shape. A worldview is never independent of the material or intellectual culture in which it is born and develops.

---

3. See Frederick Turner and chapter 1 of Hogan, *The Mind*.

There is a material reality of culture. The material culture of the nineteenth century includes trains, but this same material culture does not include jets and transatlantic flights, and the material culture of the Renaissance does not include the telephone or telegraph. Renaissance humanists thus wrote letters in Latin (the common language) to communicate with scholars in Turkey, Italy, and Spain. Intellectual culture one way or another has to accommodate itself to the realities of material culture.

This said, an important part of intellectual culture is science. Science will allow the development of many technologies, including the railroad and jet plane. So there is a mutual or a reciprocal relationship between material culture and intellectual culture. Intellectual culture finds its foundations and building blocks in material culture, but at the same time intellectual culture develops material culture as it itself develops.

There is a constant interaction and constant mutual feeding between material and intellectual culture. This mutual and constant feeding is part of what we call progress. (All those that reject the concept of progress and say that it is a European-biased, Enlightenment concept, etc., therefore deny the interaction between intellectual and material culture and condemn society to stagnation.) And there is a *constant interpretation* of both material and intellectual culture. For instance, we have science and also the philosophy of science; we have art and we have aesthetics; we have morals and we have ethics. Philosophy of science, aesthetics, and ethics are *all interpretations* of their respective material matter.

Quite often interpretations are made according to general worldviews. In this sense, a worldview would be an *arch*interpretation. If one has a religious worldview, obviously the ethics will have to correspond to this worldview. One's ethics cannot accept, for instance, adultery, and so forth. (I know you do a lot with this and categorial and practical identity.) The same goes for aesthetics. If my worldview objects to sex and nudity, well, then I will not tolerate them in a film, novel, painting, and so on.

This is true of all the arts. My acceptance or rejection of art is subject to my aesthetic opinions or judgments, and these are subject to my general worldview. This applies to all intellectual culture. We have these three categories—material culture, intellectual culture, and the interpretation of both. And we have many different sorts of disciplines devoted to the activity of interpretation: ethics, aesthetics, philosophy of science—all of these are specialized fields that engage in the interpretation of material and intellectual culture.

NARRATIVE AND POLITICS

PCH: You are raising many significant and substantial issues in these reflections. Let me select two for comment. First, let us think a little about zealots and religion. It is common to conflate them, but I think it is not quite correct. One can be a nationalist zealot or a revolutionary zealot with no less deleterious consequences (e.g., massacres of "the enemy"). Of course, sometimes people say that, in these cases, nationalism or revolution has become a religion. In one way, that is reasonable, but the result is that "religion" means something very general.

My main complaint, however, about identifying zealotry with religion can be framed in terms of the distinction you mention that I make between categorial and practical identity. Practical identity is one's activities and competences as they are engaged with the activities and competences of other people in society. Categorial identity comprises what one takes to be definitive labels for one's essence, labels for the sets that define who one is. It is well established that we form in-groups and outgroups, and thus categorial identity divisions, very easily. Even in contexts where they are temporary, such divisions can have very striking consequences. For example, they lead us to judge members of the in-group more favorably, to prefer the relative superiority of the in-group over absolute higher benefits for everyone, and so on (see Duckitt 68–69). These consequences are quite disturbing.[4]

Now, returning to religion, we might distinguish two issues. One is, say, devout Marianism or a strong devotion to, for example, the Goddess Durga. Another is a strong commitment to seeing oneself as essentially Hindu or Catholic. The former is part of one's practical identity. The latter is part of one's categorial identity. The two may co-occur. But they are largely irrelevant to one another. The sort of zealotry to which you rightly object seems to be bound up not with the devotion but with the categorial identification. That categorial identification can be found in nationalism, in ethnicity, in sex and gender, in sexual preference—in anything that can be an identity category. Whether one has feelings of devotion to a nonmaterial entity or not is largely irrelevant to this.

The second point I would like to comment on is the issue of progress. I completely agree that it is silly—well, worse than silly—to claim that the

---

4. In *Understanding Nationalism*, I outline some criteria for just when and how a label can become a categorial identity—for example, why "American" or "male" is a likely identity category for me but "Connecticut resident" or "wearer of glasses" is not.

idea of progress is Eurocentric. Indeed, I am inclined to say that there is a broad human tendency toward evaluative alignment across diverse events. This leads to two common forms of practical emplotment (especially emplotment of our individual lives or group histories). Roughly, these are "things are getting better" and "things are getting worse." Although such emplotments can be quite accurate, they can also be very distortive.

As you know, there is a strain of cognitive linguistics that tends to view our understanding of the world as guided by conceptual metaphors. Our arguments are confrontational because we use the conceptual metaphor of war to think about argument. For reasons discussed in chapter 2, this is a highly problematic view. There are undoubtedly some ways in which thinking about war or alluding to war enhances oppositional tendencies in argument. But we can choose other metaphors, such as "working together to solve a problem." In particular cases, there are undoubtedly reasons why we begin talking in terms of war. For example, if I say "Smith attacked my position," I have probably chosen the word "attacked" because I feel attacked. I probably did not start out feeling that we were working together to find a common solution, only to have my response changed by the happenstance of a cognitive metaphor. This, I think, is the value of Mark Turner's and Gilles Fauconnier's idea of conceptual integration or blending (see, for example, Fauconnier and Turner). It acknowledges that cognitive metaphors do not simply create our responses to targets. Rather, features of the metaphors are to a great extent selected by the "target" of the metaphor (e.g., argument in the case of "argument is war"). The same point holds, perhaps even more clearly, for the work of Amos Tversky and Andrew Ortony on metaphor as feature transfer and feature salience. As Ortony notes, a metaphor may function solely to make apparent some features of the target that we already knew quite well.

On the other hand, I am inclined to say that there are some ways in which our thought is often guided by very large structures. One of these, I believe, is something like an assumption of consistency of trending. It may seem that this is a transfer from physical motion and thus a form of conceptual metaphor. However, I believe that the pattern is more general. It is not just that things that are falling tend to keep falling. It is also that things that are getting old tend to keep getting old. Children who are growing tend to keep growing. I suspect that directional trending is highly abstract, perhaps akin to number, and that we are similarly genetically predisposed toward it. (On number, the reader may wish to consult Dehaene.) This cognitive tendency is exacerbated by a separate emotional propensity toward "mood-congruent processing" (see Oatley, *Best* 201 and Bower 389), our tendency to think happy thoughts when happy and sad thoughts when sad. In combination, these tend to produce telic and lapsarian emplotments—that is, "Things are getting better and better" and "Things are getting worse and worse," respectively. In fact, I think these are arguably the two most important forms of emplotment for our ordinary lives. Moreover, they tend to be self-fulfilling in that lapsarian emplotment tends to discourage efforts at improving one's conditions, while telic emplotments tend to give one confidence.

Now, back to the issue of progress. I entirely agree that we can isolate limited sorts of progress. Even so, however, we should avoid telic emplotments. Telic emplotments have two characteristics that I believe we should reject: 1) they tend to be absolute and encompassing; in other words, they tend to characterize everything as improving and as doing so consistently; and 2) in keeping with this, they tend to underestimate the value of the past.

So, I definitely agree that science makes progress. Theories generally improve. However, this does not mean that everything associated with science improves. Put simply, an improvement in theories does not entail an improvement in human well-being. Of course, there are some science-based improvements in well-being. The clearest cases of this are in medicine. On the other hand, improvements in science have also brought vastly destructive weapons systems and potentially catastrophic climate change. Moreover, the absolutist quality of science means that many promising scientific theories are forgotten—although they may be recovered at a later date.

WORLDVIEW AND IDEOLOGY

FLA: You tread cautiously here, and rightfully so, Patrick. The way I see it, as the social tissue rips more and more under the strain of an increasingly demented, violent global capitalism, the more we will need to wrap "scientific inquiry" in scare quotes. I'm still optimistic. There is still room for us to point our students in the direction of learning in the humanities that doesn't lead to solipsistic dead ends. We can lead them to generative research programs in the sciences—their findings and methods of inquiry. We can still indicate what the options carry in terms of content and in terms of their application. Hence the importance of knowing the facts discovered by science, as well as *how* all sciences proceed in their explorations of reality.

That said, let me continue with my thread: interpretation and worldview. Interpretation is done under the general umbrella of a worldview. A worldview boils down to all those rules—system of rules—that will be normative in your interpretive activities; that is, they will be acting as rules in our interpretive activity; they are intellectual choices that will function as norms within our interpretive activity. So a religious worldview that excludes chance and affirms that everything that happens to the individual is preordained will function as a norm in the interpretative activity of the person who holds such a worldview.

Quite generally interpretive activities are realized or take place under the guidance of norms that constitute (or conform to) what we call a worldview. Some have a philosophically idealist worldview, others a materialist; some have an atheist and others a religious worldview.

In narrative fiction, a specific domain of aesthetic activity, an author creates a work of art according to aesthetic goals and aesthetic aims, and these goals and aims are *ruled* so to speak by the specific worldview of the artist concerned. However, an artist's worldview does not necessarily determine his or her aesthetic aims. For example, some scholars consider Balzac's novels to be politically reactionary because politically he considered himself in general terms to be a monarchist and not a republican. But what we find in his novels is very different. His careful detailing of all walks of life—the squalor and sordidness of a capitalist society—and the creation of all variety of situations and relationships between his characters creates a rich, whole world. That is, the image that is generated and the fictional world that is created is not a hymn to monarchy. I do not turn the last page of *Le père Goriot* (1835) and think "We need a monarchy."

My point is that an author can have a worldview that is opposite to his or her aesthetic goals. For instance, Balzac's aesthetic goals were to invent characters in situations and relations among characters that would *represent* what in Balzac's time and in his mind were called "social types": the miser, the stock-market speculator, the good grandfather, the bad child, peasant, and so on. He called the totality of his work his "la comédie humaine"—narrative fictions that covered all sorts of types that are found in society. This aesthetic goal together with the aesthetic means that we call realism in fact clashed with the goals of his personal worldview. Balzac's aesthetic aims had such a central importance in his everyday life that he couldn't do away with them. They overruled his personal interpretive worldview aims. Every time he sat down to write, it was his will to style that overpowered him.

We also have the figure of Mario Vargas Llosa, who has maintained a very clear sense of himself as a writer free to explore any and all things under the sun—and beyond. He has written many political articles, and he campaigned to be president of Peru in 1990. But he is acutely aware of the dangers awaiting any writer of fiction who starts believing that fiction has the powers that in fact pertain only to politics. In a 1991 interview with Susannah Hunnewell and Ricardo Augusto Setti for the *Paris Review,* he sums this up nicely: "I think it is important that writers participate, make judgments, and intervene, but also that they not let politics invade and destroy the literary sphere, the writers' creative domain. When that happens, it kills the writer, making him nothing more than a propagandist" (http://www.theparisreview.org/interviews/2280/the-art-of-fiction-no-120-mario-vargas-llosa).

Many books by Vargas Llosa deal directly or indirectly with politics. But he keeps his political beliefs at arm's length in his attempt at writing the total (ideal, inaccessible) novel. It is this totalizing and nonsentimental approach to fiction making, already posited in his twenties, that led him to a great payoff. The publication of his first book in 1959, the short story collection titled *Los jefes* (titled *The Cubs and Other Stories* in the 1979 English translation), along with the 1963 publication of *La ciudad y los perros* in 1963 (titled *Time of the Hero* in the 1966 English translation) earned Vargas Llosa a whirlwind of laudatory praise; the latter publication earned him the prestigious Premio de la Crítica Española award.

In contrast to Vargas Llosa with his admonition to keep separate political leanings and activities as a citizen from fiction writing, we have the figure of John Dos Passos. At a certain moment in life, his fanatical rejection of communism (he erroneously conflated communism with the Stalinist, counterrevolutionary bureaucracy) and zealous, hawkish procapitalism penetrated his aesthetic through and through. The radical change in his worldview led to the abandonment of many of his aesthetic motivations and goals. Unable to keep at bay his fervent procapitalist worldview, he wrote *Brazil on the Move* (1963)—a tourist guide that doubles as a reactionary eulogy to capitalism in the form of a Brazilian tobacco company—and followed this critical and mainstream failure with *Century's Ebb* (1970)—an equally dull, throwaway product.

PCH: You are getting at something very important here, though I would put it somewhat differently. First, I initially balked at your references to worldview. I am not sure I believe that people generally have worldviews. But as you have developed your idea, I see that there is more convergence between our positions than I initially assumed.

Here is the model I would use. We all have attitudes, preferences, and the like. Some of these are stable, while others vary circumstantially. The configuration of such feelings is not necessarily consistent across time—or even at a given time, as the experience of ambivalence attests. These attitudes and so forth have many sources, but their consequential operation is largely in subcortical emotion/motivation systems. We also have various complexes of ideas, expectations, beliefs, and so on. Like the emotional preferences, these may be enduring or contingent and they need not be consistent, even at one time. These largely bear on prefrontal cognitive systems. Both emotional and cognitive systems have conscious or explicit and nonconscious or implicit components. The implicit components are necessarily far more numerous than the conscious components and far less open to self-conscious control.

What you are calling "worldview" seems to me to fall largely under the category of explicit, enduring, cognitive features—thus, for example, self-conscious beliefs. I think we may diverge some on the importance of worldview understood in this way. (Of course, you might also simply reject construing "worldview" in this manner.) In my view, explicit, enduring, cognitive commitments play only a very limited role in our behavior. Specifically, our actions—in a very broad sense of undertakings, including interpretations—are initiated by largely contingent and partially implicit motivations, then developed by largely contingent and partially implicit cognitions. As we engage in actions, we sometimes become aware of contradictions with explicit beliefs, long-term goals, or other self-conscious commitments. That is the point at which worldview enters. At that point, we need to either change our behavior to make it consistent with the explicit beliefs, and so forth, or we need to reinterpret the beliefs to make them consistent with the behavior, or we need to somehow keep the explicit beliefs out of our minds. By this account, worldview often has no bearing on behavior, and, when it does have bearing, it is more equivocal and ambiguous than we are usually inclined to believe.

Of course, worldview—or ideological commitments—may have a very significant impact on what one says and how one judges others. When Jones's own behavior is at issue, then the key factors are Jones's implicit motivations and cognitions. But when Jones sees Smith doing something, then the crucial factors for Jones are likely to be Jones's explicit cognitions. In other words, if Jones has a strong commitment to some religious or political ideology, there is some likelihood that he will apply explicit standards more rigorously to Smith than to himself. Thus this model predicts that there should be some tendency toward at least apparent hypocrisy on the part of all of us.

Now, let us return once more to simulation. When an author engages in simulation, he or she will proceed predominantly through implicit motivations and cognitions. Self-conscious worldview or ideology will tend to enter only at points where some contradiction becomes salient. That is why, as Lukács famously argues, Balzac is not confined to his class ideology. Specifically, citing Engels, he praises "the correct and profound reflection of reality which rose above Balzac's own individual and class prejudices" ("Marx" 138). In the terms I have been using—which contradict some of the implications suggested by "reflection of reality"— Balzac's simulation is relatively unaffected by his explicit cognitions and affiliations.

This also indicates why it is often problematic when authors follow through on some self-conscious political program in their writing. In fact, I would say that the problem is general and applies not only to the incorporation of a political view but also to systematic incorporations of an ethical view or a psychology (e.g., psychoanalysis—or current neuroscience). Our self-conscious explanations tend to be much more unequivocal, much less motivationally and cognitively complex, than our simulations. The result is that "programmatic" works may become mechanical and simplistic.

On the other hand, it can happen that an author's comprehension of a theory is itself complex, so that it adds nuance to his or her simulations rather than substituting for them. I believe that this at least sometimes happens with Brecht, for example, in his use of Marxist thought in, say, *Die Maßnahme* (*The Measures Taken*). Another case might be M. F. Husain's use of Sufism in *Meenaxi*. In both these cases, though, the self-conscious worldview or ideology enhances the insight and aesthetic effect of the work precisely insofar as it does not displace spontaneous simulation but rather expands the possibilities for such simulation. For example, Brecht illustrates the problems with canonical errors condemned in orthodox Marxism (e.g., voluntarism). But at the same time he simulates the complexities—thus insights and values—of the erring (voluntarist) revolutionary, as well as the problematically reductive judgments and cruel actions of the party leaders.

FLA: With this in mind, Fuentes enhances his narrative fiction (your simulation) because of a conscious awareness of a worldview; the latter Dos Passos diminishes and turns mechanical his narrative fiction (your simulation) because he fails to hold his worldview at arm's length. I believe we're on the same page here. Let me dig at something deeper: the domain of culture we call interpretation. As I mentioned, this domain possesses as a general umbrella a worldview from which interpretation derives or takes place; this worldview will inflect interpretation in one direction or another. At the same time, there is no *determinism* such that the worldview will necessarily determine in fashions A, B, C, interpretations of W, Y, Z. There is no fatalism; there is not a necessary one-to-one correspondence. Interpretations are guided and ruled in a more or less flexible way (depending on the individual) by a worldview.

    In the reading of fiction, this can lead to a conflict of interpretation of worldview between author and reader. An author sets aesthetic goals and means that he or she interprets as pertaining to his or her worldview (atheist), but it could happen that the reader fills in the blanks in the blueprint according to a religious worldview and so reads the work with an interpretive key of his or her religious worldview. Since there is total freedom in terms of the interpretive activity of the author and in setting goals and means, it could happen that the equivalent features of freedom in the reader lead to radically opposite interpretations based on contrasting worldviews. This type of symptomatic reading happens all the time when I teach, by the way.

PCH: That certainly makes sense. Here, a set of distinctions might be useful. There is spontaneous, implicit interpretation—what hermeneuticians refer to as "understanding." Then there is effortful, explicit interpretation—what hermeneuticians refer to as "explication." Then there are the various conditions in which these interpretations occur—what hermeneuticians refer to as "application." When Jones tacitly reinterprets his religious beliefs (e.g., "Thou shalt not kill") in relation to his emotional support for war, that is different from an undergraduate student being asked to come up with an explicit, effortful interpretation of Faulkner in a class, and that is in turn different from a graduate student or professor trying to produce an interpretation of Faulkner that will get published. You are probably right that worldview or ideology may enter in the second case (the student), given certain further conditions. The first case goes in the opposite direction, as discussed earlier. The third, professional case appears different from both, though it is likely to involve bad faith. It seems all too often to concern not a worldview or ideology that the interpreter holds or tries to reinterpret. Rather, it manifests a set of imitative procedures that the interpreter imagines evaluators (e.g., journal referees) will judge positively.

ETHICS AND EMOTION

FLA: You mention bad faith, reminding me to pick up the thread of ethics. Narrative fiction deals mostly with humans or humanlike entities and with their behavior and the circumstances in which they live and therefore relate to one another. That is, in general, almost as a rule (some people consider it a universal rule with no exceptions, but I am not sure about this), stories are about humans, their circumstances, their behavior, and their reciprocal relations. It is to this material that the generative operator of discourse is applied.

So the many features of humans become fictionalized by the application of this generative operator; they thereby cease being human and become creations, creatures, constructs. The same applies to circumstances, behavior, and relations. When this happens, we have fictive behavior, for instance, and therefore we have *fictive morals;* and we can have at an upper level *fictive ethics*. That is, in narrative fiction we have fictive rules applying to characters: good and bad behavior are the two categories, and at the upper level, fictive ethics has as its main ruling concepts the concepts of good and bad—and, of course, the concept of prohibition and punishment.

Since in real life, real interactions between individuals are so important, we have all kinds of rules governing this behavior—good versus bad manners, for example—and rules governing morals—thou shalt not X, Y, or Z. In real life, since we all live in a social setting and to be human is to be a social being (we cannot survive without social structure), morals and its interpretation (ethics) have a very relevant, very foregrounded important place.

When all these ingredients are fictionalized by the generative operator of discourse in a story, we also centrally have in our narrative fiction—in a very important way—fictionalized morals and a fictionalized ethics ascribed to our fictional characters.

An important part of the work of the artist creating narrative fiction is *to write in the diegesis* (which is different from story; there are stories about real events—anecdotes, for instance—and we apply the operator of discourse to these real events). Within the diegesis the author creates characters and relationships, and a fictional morality rules the relationships between characters. And, of course, the author creates all the other stuff of diegesis, such as setting, circumstances, and time and place.

PCH: On the issue of whether stories always concern agents—this is largely a matter of where one wants to draw the dividing line. In other words, it is definitional. A famous example is the exploding radiator, discussed by the philosopher Carl Hempel in the course of a consideration of laws and events in relation to the study of history. George Reisch comments that Hempel's "explanation of the exploded radiator, after all, is a mini-narrative: 'Once upon a time, there was a radiator. Then, it got really cold and the radiator burst because it couldn't withstand the pressure exerted by the freezing water it contained'" (19). So, is Hempel's account of the exploding radiator a story? If we want, we can say it is—or not.

FLA: Perhaps this is where ethics and its necessary emotion correlates does allow for the definitional. The rules governing the relationship of characters—this could be your exploding radiator or a radiator hose or Cortázar's otherworldly creatures such as his *cronopios* and *famas* or anything with agency, intentionality, sense of causality, and movement—within the diegesis are also a fictive ethics. This strictly applies mutatis mutandis to emotions. An important part of our lives is emotion driven and ruled. The emotions are the air we breathe. They are a very important part of our relations with others, whether in the form of love and hate or indifference and neglect (passive-aggressive forms of behavior).

Emotions have such a central importance in the everyday lives of real people in real time and space that they naturally appear also as inspirations for authors. All the emotions of the husband and the lovers of the real woman who committed suicide informed the making of the real story that Flaubert uses for *Madame Bovary;* when he submits this real story to the treatment of fiction, to the generative operator discourse, what happens is that a very important part of the work of the artist becomes finding ways to ascribe emotions to characters and finding ways and means to show these emotions so that the reader can recognize them as being reliable or credible or, one way or another, can recognize them as emotions.

What I'd like to insist on is that emotions in narrative fiction are *constructs*. I deliberately decide that this character as my creation will fall in love with this other character or will hate this other character.

Emotions in fiction are *fictional*. The scientist who decides to study emotions scientifically in narrative fiction texts will necessarily do a very bad job. Whatever emotions are in the text are there because the author put them there; therefore, they are already *known as emotions*. They are a part of the prior knowledge of the author; they are part of the story that exists in advance of being submitted to the fictional, specifically generative discourse, procedures.

PCH: I would once again say that the emotions are simulated rather than constructed. That may be a simple preference in words. But "constructed" makes it sound as if the process is more self-conscious than simulation is. It also suggests a difference in process between one's imagination of characters and one's imagination of real people. I do not believe there is a difference, at least not a difference in kind. That is why I argue in *What Literature Teaches Us About Emotion* that studying emotions in narrative fiction should be part of affective science.

FLA: Patrick, I'm not sure I was clear. Indeed, I use constructed to underscore my understanding of narrative fiction to be a newly built object in the world that requires the conscious will and action of the creator in its making. Of course, many more authors than not (and we could say the same of musicians, architects, artists, and the like) are *not* willful creators. For every thousandth building we drive by every day, maybe only one stands out as having the presence of a will to style, say. I like that we leave this part of the conversation with a nod toward your book and more generally your work. I believe our work, while differently orientated, offers options for pursuing research programs that aim to formulate a foundational study of narrative fiction specifically and aesthetics generally—the subject of our next section.

CHAPTER 4

A Scientific Approach to Aesthetics

FREDERICK LUIS ALDAMA: Before going into the specific discussions of literature, and the equal validity of interpretations, some general remarks are in order.

First, even assuming the truth of the most radical ascriptions of a modal architecture of the brain, and even assuming that some areas of the brain are "dormant" while others are active in such and such a moment, I still believe that the brain is a unified whole. It is not a sum of separate and independent modules.

The brain works as a unit. When I am contemplating a painting or reading a novel, it is not just a few areas of the brain that are engaged—the language area for instance, while the whole limbic system is inactive. Rather, there is a constant interaction between the neocortex and the limbic system; and within the limbic system there are areas such as the hippocampus that fire more intensively. We activate the cognitive, the linguistic, the limbic system (emotion system), and the perception and memory systems when reading a novel or contemplating a painting.

Second, all these neurobiological functions operate at the same time as *meaning-conferring mechanisms*. For instance, perception is not just the sensory grasp of external reality. Perception is a faculty that operates as a shape- and meaning-conferring mechanism. When I perceive an object I do not just have a physiochemical reaction. In addition, my eyesight turns those sensorial inputs into perception by conferring shape, by isolating objects from each other.

Perception is a shape-conferring neurobiological function. And, by the same token, because it confers shape, it also confers meaning. All aesthetic experience is about shape and meaning. There is no shapeless aesthetic experience; there is no meaningless aesthetic experience. This shape could be a house in architecture, movement in ballet, colors in a painting, and so forth. Shape and meaning come together in the act of perception. This is why we can easily say that perception is a shape- and meaning-conferring mechanism.

Third, from a neurobiological point of view, the two main essential and perhaps, according to some scholars, the "only" senses involved in the aesthetic experience are vision and hearing. And that's because the olfactory, the gustatory, and the tactile senses have a very low capacity—perhaps a zero capacity—to confer shape. These senses are more remote from our most intense common experiences that are based on the visual and aural senses.

This is not to say that one cannot train these senses to give shape to smell, taste, and touch. There are those who educate the olfactory sense and train it, so they can delimit the line between one odor and another in the making of perfume. And others train their taste buds to delimit the boundaries between different tastes to evaluate wine (origin, ingredients) and also to develop new combinations in the making of wine. Great chefs like my sister likewise have educated their tastes to be able to tell you the ingredients of food and even the respective amounts of the ingredients. In a much more occasional way, the tactile sense is used in contemporary forms of art in which the artist uses paints and collage devices to create a surface texture. In the making, artists use materials for the collage that feel right to their fingertips that they then apply to the surface of the canvas. The paint itself can also leap from the canvas, eliciting an aesthetic response from the audience that triggers that impulse to want to touch its surface with the fingertips. Sculpting, of course, while centrally a visual aesthetic experience in creation and reception, is also a tactile experience.

Much contemporary poetry aims at creating an aesthetic experience through the simultaneous use of the shaping or the patterning of sounds in the poem *and* the visual presentation of the poem on the page. A poem can be read and enjoyed in terms of its sound patterns and simultaneously enjoyed visually, because those sound patterns are distributed in the form of a spiral. Urayoán Noel's collection *Hi-Density Politics* (2010) is an excellent example of contemporary Latino poetry that does this to great effect.

Technical devices in filmic narrative fiction are also a form of segmentivity charged with meaning because the cuts give specific shape to each sequence of frames. We also see this at work in comic books. In a much looser way, we see this in the distribution of the text in a short story or novel. In *The Complete Western Stories of Elmore Leonard* (2007), we see not only a careful use of crosscutting between one paragraph and another to increase the dynamism of the story but also a use of cutting within a story that makes it seem as if we were encountering a new chapter within a given short story. That is, Leonard segments an already rather brief story by introducing a chapter heading within it. This creates a dynamic reading experience that actively involves our perception (visual experience) of the segmentation of the text.

AESTHETIC PLEASURE

PATRICK COLM HOGAN: You raise many interesting and significant issues here. For example, you suggest, if I understand correctly, that shape is a necessary condition for aesthetic pleasure. The first question here would be—what constitutes shape? You seem to have in mind a figure in a figure–ground relation or something along those lines. That is, at least, what "shape" suggests. On the other hand, it seems that we can have beautiful backgrounds, even such things as a beautiful sky or an aesthetically pleasing color.

I think you are on to something here, but it might be worth backing up for a moment and considering other issues. First, there is the issue of sensory modality. We of course refer to visual art and visual scenes, as well as music and sounds, as beautiful. We probably do not refer to any touches as beautiful. But we do think of some textures as intrinsically pleasing. So I might remark that a piece of velvet has a lovely plushness, urging you to touch it also. The second point seems relevant because we often have a strong inclination to share aesthetic pleasures with friends. Note that the pleasure of touch in this case is not contingent on any further, nonaesthetic pleasures. More significantly, it is clear that the somatosensory cortex is involved in our aesthetic response to dance (see Calvo-Merino and colleagues). Thus it seems clear that the somatosensory system has aesthetic properties as well. We do not typically refer to a beautiful smell, but we do refer to a "lovely bouquet," which conveys much the same idea. The difficulty with taste is that it is routinely bound up with hunger satisfaction. However, the system of taste preference is in fact distinct from that of hunger satisfaction. (We discuss this further in the next chapter.) Thus we do often undertake to experience taste pleasure apart from hunger satisfaction. Wine tasting, which you mention, is an obvious case. Moreover, taste serves as a recurring model for other modes of aesthetic pleasure. For example, the central concept in the Sanskrit aesthetic tradition is *rasa,* which "in its most literal sense means juice, taste, flavor." In aesthetics, it is used "to express the flavor or mood which characterizes a play" or other work of art (Ingalls 15–16).

So, we may disagree as to whether aesthetic experience is confined to particular sensory modalities. The question that arises here is whether all these cases are really instances of the same sort of thing. In other words, is aesthetic pleasure in wine tasting the same sort of thing as in reading poetry, or are we simply being misled by some similarities in phrasing? Note that the same point holds for reading poetry versus looking at paintings or looking at landscapes or hearing music, and so on.

There are several ways in which we could seek such continuity. These fall into two broad categories: continuities of beauty and continuities of aesthetic experience. Continuities in beauty would comprise recurring properties in the objects—for instance, shape, in your account. Continuities in aesthetic experience would comprise recurring properties in response; your reference to the experience of meaningfulness may be an instance of this.

There are two obvious ways continuities in experience could occur neuropsychologically: cognitively and affectively. Cognitive aesthetic continuity would involve more abstract, amodal representations of aesthetic objects. For example, there appear to be neurons in the parietal cortex that code for number independent of modality (see Dehaene). If the same were true of aesthetic experience, that would provide evidence of an amodal sense of beauty. Moreover, in that case, the existence or nonexistence of an olfactory sense of beauty would be an empirical, neurological question. Affective aesthetic continuity would be secured by some emotion system or complex of emotion systems that recur with different aesthetic experiences. Whether there is indeed such recurrence is likewise an empirical, neurological question.

Again, we may alternatively wish to seek for aesthetic continuity in the objects themselves. Our neurology may or may not be the same for, say, beautiful music and beautiful painting. Either way, there may be properties shared by the objects themselves.

My inclination is to view "the aesthetic" as a complex, prototype-based concept. We may isolate different elements that contribute to an account of something being aesthetic—both objective and experiential elements. The result is that some objects are better cases of the aesthetic than others. I do not necessarily mean that they are more beautiful. I mean that there are some objects that fulfill more of the criteria for being aesthetic. Indeed, given what I have just said, perhaps it would be more appropriate to say that some *events*, comprising an interaction of a subject and an object (or situation), satisfy more criteria for being aesthetic.

Take, for a moment, the emotional part of aesthetic experience. The first thing that we would be likely to say about aesthetic feeling is that it must involve sustained interest. Thus it must to some degree activate attention systems. Moreover, in keeping with this, it must not be habitual. Habituation reduces attentional orientation (Frijda 318) and emotional responsiveness (LeDoux, *Synaptic* 138). This is, broadly, a point stressed by the Russian formalists, though in a very different theoretical context (see Shklovsky 741). It is also suggested by Kant (*Critique of Judgement* 80).

Beyond attentional engagement, we would expect some sort of reward involvement. This is not to say that there would be no aversive emotion. However, there must be some enjoyment for the experience to count as aesthetic pleasure, and there is neurological evidence that aesthetic preference is connected with reward systems. Thus Hideaki Kawabata and Semir Zeki ask what conditions are "implied by the existence of the phenomenon of beauty (or its absence)." They conclude that "in esthetics, the answer . . . must be an activation of the brain's reward system with a certain intensity" (1704). Kawabata and Zeki are concerned primarily with visual art. Peter Vuust and Morten Kringelbach report research on the importance of the reward system in music. In his more general overview of aesthetic preference, Martin Skov points to the centrality of "the processing of reward" (280).

Interest and pleasure may be necessary conditions for something to count as an aesthetic experience. I would add that an experience becomes more prototypical when other emotions are present as well. Of these, perhaps the most important and consequential is some form of attachment feeling. We see this most obviously in judgments that one's beloved is beautiful. It is a commonplace that one's experience of one's beloved is an aesthetic experience. In a well-known poem, Sappho writes about how different people give different accounts of what is most beautiful. But, in her view, it is "whatever one loves" (41). There is, in fact, neurological support for this, though it has not been widely understood in aesthetic terms. Specifically, Nadal and colleagues point out that some research has "found lower preference ratings associated with decreased activity of the caudate nucleus" (388; see Vartanian and Goel). The point is important because of our more general understanding of the caudate nucleus. Specifically, seeing the face of an attachment figure has been linked with "activity in the caudate, a deep brain structure associated with feelings of love" (Arsalidou and colleagues 47). The "caudate . . . has been associated with feelings of maternal and romantic love" (50). Similarly, in Jaime Villablanca's account, "the specific function of the caudate nucleus is to control approach-attachment behavior," including "romantic love" (95). This is why, for example, Proust can refer to Swann as being "in love" with the "little phrase" of Vinteuil's sonata (255).

FLA: Yes, a willful shaping is a necessary condition for our interest and pleasure in the experience of the object as well as crucially in our meaning-making operations. As we discussed, this can involve the syntactic shape or form of the phrase as well as segmentation and a myriad of other generative devices (or tools) . . .

PCH: That leads back to the issue of beauty or object properties. Here, there appear to be two different ways in which object properties bear on aesthetic events. Of course, neither is independent of subjective conditions. We need to encode the relevant properties and process them appropriately. (To encode properties is to select them from the perceptual environment, organize them into units, and bring them into structural relations with one another.) If we do not encode the properties they have no existence for us and thus no effect on us.

The two types of properties I have in mind bear on, roughly, objects as such and complexes of objects—with "object" understood very broadly as any cognitive target. Alternatively, we might conceive of the difference as parallel to that between the ventral and dorsal streams of visual processing—that is to say, the "what" and "where" streams. As Robert Wurtz and Eric Kandel explain, "motion" is "mediated in large part by the dorsal pathway," while "contours" are "mediated largely by the ventral pathway" (548). In connection with these divisions, one type of beauty concerns intrinsic properties and another concerns relations across objects, locations, or moments. (V. S. Ramachandran too sees the dorsal/ventral division as important to differentiating aesthetic criteria, though his specification of those criteria is very different; see his chapter 8.)

Your idea of shape relates most obviously to the former variety. As you have phrased it, it seems perhaps related to Zeki's idea that art seeks to isolate "the constant, lasting, essential and enduring features of objects, surfaces, faces, situations, and so on" ("Art" 4). Zeki relates this to the brain's operation in isolating constancy in the context of continual change.

Zeki cites many artists and many works of art for which this account initially appears fitting. However, I believe it is mistaken. For example, in citing artists' explicit statements about art, Zeki seems to stress artists who are reacting against impressionism, with its emphasis on the fleeting quality of experience. The rhetoric of these anti-impressionists accommodates Zeki's hypothesis but for historically contingent reasons. Moreover, the history of perspective painting likewise seems to contradict Zeki's view. Of course, Zeki does treat artists such as Monet. But his interpretations are biased by his theoretical presumptions. For example, he takes Monet's paintings of "the same scene in differing conditions" as evidence that Monet was "search[ing] for constancies" (*Inner Vision* 210). The more obvious interpretation is that he was searching for differences, the differences in the changing conditions that would be highlighted by the presupposed (thus not sought) constancy of the object (e.g., the Rouen Cathedral).

On the other hand, I believe Zeki is on to something. Beauty is not so much what is constant, but, so to speak, what is "normal"—and the normal is, of course, related to the constant. Specifically, we may take a clue from research on facial beauty. Judgments of facial beauty seem to be governed by a principle of averaging, such that the most average face is the most beautiful face (Langlois and Roggman). The point, though not precisely predictable, is also not surprising in light of how our processes of categorization appear to work. Specifically, we seem to understand categories by reference to prototypes (see Rosch). Prototypes are, roughly, weighted averages across experiences. For example, we categorize faces as faces. We then form a prototype of a face by averaging across the different faces that we see, probably with some degree of difference depending on how long we see a given face, how thoroughly we encode the features, and other factors. This averaging appears to be "weighted" in the sense that it will be slightly different from a statistical averaging, as a result of contrast with opposed categories. Thus our prototypical man may have more "masculine" features than the statistically average man, due to a contrast with "woman."

The process of weighted averaging in itself might not lead us to expect that prototypes would activate some reward response. However, once the possibility is suggested, then the idea that it would occur for evolutionary reasons seems quite plausible. Steven Pinker points out that we prefer symmetrical faces and notes that this has reproductive benefits (484). The general idea seems completely correct. However, the case is too particular. Specifically, we would expect prototypical instances to generally be less potentially problematic (e.g., less likely to be diseased), whether we are speaking of animals or fruit or faces. Thus it seems unlikely that there is a specific adaptation for facial symmetry preference. Rather, it seems likely that there is an adaptation for seeking proximity of the target to the relevant prototype, whether that is the prototype of a face or the prototype of a carrot. This seems likely because prototypes would tend to eliminate risky deviations from normalcy across the board, for carrots as much as faces. Obviously, like any other evolutionary mechanism, this means that it will often get things wrong and lead us to disprefer something that would in fact be better for us. The point is merely that, on the whole, the mechanism should approximate the function of, say, avoiding disease or similar harms.

At this point, the question arises as to whether facial beauty is really a matter of prototype approximation. It seems that it is. The point appears to be supported by studies bearing on the contrastive operation of prototypes. Specifically, by a prototype account, we would expect judgments of female facial beauty to be enhanced by more distinctively female features and for judgments of male facial beauty to be enhanced by more distinctively male features. In keeping with this, Russell reports research regarding male/female luminance differences around the mouth and eyes. As Russell summarizes, "Female faces were found to be more attractive when ... luminance difference was increased." This is because "the luminance difference between the eyes and mouth and the rest of the face is naturally greater in women than men" (1093). Thus the preference is contrastive in the way that we would expect. Of course, we cannot simply assume that this is a matter of beauty, since sexual desire may be involved (see Franklin and Adams on the complexity of judgments regarding facial attractiveness). Moreover, Pinker's suggestions about testosterone levels and gender differences in facial structure seem likely to have bearing here. But the preference is also consistent with the predictions a prototype account would entail. Indeed, one would expect sex-based prototype weighting for beauty and sexual attraction to interact in relevant cases, just as attachment and sexual desire systems appear to interact in romantic love (see, for example, Shaver and Hazan 482).

Here, then, one might ask if prototype-based aesthetic judgments are confined to faces. In fact, it seems well established that they extend well beyond faces. For example, Colin Martindale and Kathleen Moore provide considerable evidence that, for color, "more prototypical stimuli" are "strongly preferred over less prototypical stimuli" (670). There would presumably be some qualification for this relative to habituation. Nonetheless, there is evidence that prototypes affect aesthetic preferences in such apparently habituation-prone areas as furniture choices (see Whitfield and Slatter).

In short, it seems that prototype approximation plays an important role in defining beauty and aesthetic pleasure. Even so, it is clear that prototypicality does not cover all cases. The other criterion for beauty seems to be (unexpected) patterning. Some neuroscientists stress the pattern-isolating function of the brain. Indeed, Gerald Edelman goes so far as to claim that the "primary mode" of thought is "pattern recognition" (103), that "brains operate prima facie not by logic but rather by pattern recognition" (58). This is presumably the result of evolution, with pattern recognition having adaptive benefits. In keeping with this, we might expect the isolation of patterns to yield endogenous rewards, thus motivating pattern-seeking behavior. This would explain some aspects of, for example, musical beauty in which we isolate sequential patterns and transformations of those patterns in thematic variations, key modulations, and so on.

It is worth pausing for a moment over the case of music. Vuust and Kringelbach present considerable evidence that pleasure in music is bound up with "anticipation/prediction" (256). Indeed, they directly link anticipation/prediction to reward-system engagement (256, 266). Vuust and Kringelbach stress the pleasure of getting the prediction right. Indeed, they see this as having adaptive consequences since an endogenous system "rewarding correct predictions" will "reinforce correct predictions of the future" (266). Unfortunately, this does not quite fit their own data. Specifically, they explain some of the most intense experiences of musical pleasure as resulting from "failure to predict future events" (262) and thus as involving "a violation of expectancies" (263). Clearly, expectancy is the isolation of some temporal pattern. It would seem that a violation of expectancies is simply a disruption of the pattern. However, it is likely that intense musical pleasure does not result from all violations of expectancies. Rather, it seems likely that it results from a violation of expectancies that the listener can quickly accommodate into another pattern. The violation works against habituation. But the subsequent recognition of a pattern yields the reward.

Indeed, this analysis fits their evolutionary model with only a slight alteration. Specifically, we would expect pleasure to result from getting a prediction right, as they say. But that pleasure should habituate with time. More importantly, if we have evolutionary reason to enjoy true predictions, we have, if anything, even greater evolutionary reason to learn new patterns. That is, after all, how we come to make (subsequent) predictions in the first place. Thus there should be reward involvement for learning a pattern after a false prediction.

Beauty in literature appears to combine both patterns and prototypes. Moreover, it does so at different levels. For example, we isolate patterns in speech rhythms (e.g., in poetry) and in imagery. The praise of the beloved's beauty in love poetry involves prototype development. Emotion-system contributions enter here as well. For example, literature treating romantic love or parent–child relations—and thus a good deal of literature worldwide—is likely to activate emotional memories connected with attachment.

In saying all this, I am leaving aside a complication: prototypes seem to come in different forms. Some are more governed by the "mean," others by a contrastive "limit."[1] Thus our prototypical diet food is not an average of diet foods, not even a weighted average. It is an extreme or limit case—the zero-calorie lettuce (Kahneman and Miller 143). But that too raises interesting questions concerning the degree to which the contrastive weighting of prototypes governs aesthetic preference. V. S. Ramachandran has maintained that beauty is bound up with the "peak-shift" effect. The peak shift occurs when an animal has been trained to respond differently to two distinct stimuli, say a 2" × 2" square and a 2" × 3" rectangle. Through training, the animal's peak response is to the 2" × 3" rectangle. However, if faced with, say, a 2" × 4" rectangle, the animal's peak response shifts to the new stimulus, which exaggerates the difference between the (unrewarded) square and the (rewarded) 2" × 3" rectangle. (Ramachandran discusses the point in detail in chapter 7 of *The Tell-Tale Brain*.)

Ramachandran does not report whether there are limits to this, so it may or may not involve limit prototyping. Other cases cited by Ramachandran almost certainly are a matter of weighted averaging. For example,

---

1. See Hogan, *Affective* 190.

Ramachandran suggests that the aesthetically ideal female form exaggerates distinctive female features. But there is presumably some point at which the figure's waist becomes too slender and the breasts become too large (e.g., I doubt that people would judge a woman with a 60" bust and a 1" waist to have a singularly beautiful physique). Presumably, the same point would hold for male beauty. In any case, Ramachandran's observations do suggest that there is a complex interaction between statistical averaging and contrastive weighting and that the best account of beauty would need to treat this.

SOURCES OF AESTHETICS

FLA: You cover much territory here, Patrick. I like very much your clear explanation of interest and pleasure as they relate to the learning of new patterns. You mention love poetry. I would like to take us not so much exclusively in the direction of romance, but in that of human relationships and subject-to-object relations more generally. In very broad outlines and with very broad brushstrokes we can distinguish within human behavior four kinds of relationships: 1) between humans and nature; 2) between humans and the tools or instruments they create to modify nature; 3) between one human and other humans, between the human and his or her society; 4) between humans and nature *and* tools *and* other humans from an aesthetic point of view. It is important to make these seemingly artificial and general distinctions to begin to outline the scientific criteria for aesthetics.

We have had a specific relationship with nature: it has been the source of the means of survival since our hunting and gathering days and before that, as paleoanthropologists have discovered, as essentially scavengers. With the cutting of rock for spears and other tools (which enabled humans to kill animals themselves, thus eliminating the need to have to wait for another animal to finish eating its prey and then move in to get whatever remained) and then the ability to produce fire, the relationship to nature changes. So here we also have a relationship with the tools produced in the transformation of nature. Then we see the primitive origins of morals: the creation of rules that will govern the relationship between one human and other humans. This gives rise to ethics and to other forms of relational behavior such as magic and religion.

Humans find, finally, beauty and ugliness in nature *in* the tools that they make and *in* the relationships among humans. This is what we call the *aesthetic* relations; that is, relations based on these emotional reactions we call beauty and ugliness.

All these relations and their effects and consequences in terms of institutions and rule-governed social life—aesthetics, politics, laws, ethics, altruism, protection of offspring, and so forth—form an increasing number of separate, independent domains. Nowadays, human relations in society that are the object of specific rules have given birth to things such as constitutions and courts (starting with a high court, followed by a series of other levels that specialize in different kinds of relations, labor, business, patent, etc.). We have a myriad of laws and law books and specialists in law. Then to ensure that the law is obeyed, besides the courts we have all kinds of specialized people who are trained to hurt, maim, detain, restrain, damage, attack other people (this is function of the police and army). Then we have specialized centers that detain people: jails, prisons, concentration camps, and so on.

These are just a tiny example of the complex web of relations that have their origin in the Paleolithic. I could give many other examples of how these forms of human relationships have evolved and become more diverse and specialized.

It is the same with aesthetic relations. It is likely that, for instance, the Paleolithic hunters making their weapons would have taken a certain pride in a work well done. That is, I would feel more satisfied with the result of work if my arrowheads were sturdy and at the same time lighter than other arrowheads and therefore perhaps could be shot farther away from the prey and have the same effect as heavier ones launched closer to the prey. This would give me a feeling of satisfaction. So, I would take an ever-increasing satisfaction in improving on the results of my work. My skill and virtuosity as well as the communicating of this to the next generation all inform my satisfaction in the making of an innovative arrowhead.

PCH: The human relations aspect here of course returns us to the issue of attachment. But I'd like to pick up on a different aspect of what you just mentioned—ethics. Ethics is relevant here for a number of reasons. I would like to focus a bit on one. We sometimes experience abstract objects as beautiful. Thus, for example, aspects of mathematics are beautiful. An important case of this is, precisely, ethics. One of Kant's most famous statements, made in a nonaesthetic context, is that the feeling of the sublime is inspired most profoundly by the starry sky above and the moral law within (*Critique of Practical Reason* 166). The starry sky points toward our material insignificance, he explains, while the moral law reveals our unbounded possibilities. In the case of mathematics, aesthetic feeling is at least in part a matter of unexpected pattern isolation. There is some element of that in the starry sky, and some element of prototype approximation too. But the applicability of these ideas to the night sky is less clear than to the blue sky of day. More significantly, it is not evident how these criteria might apply to the moral law. Admittedly, there is some element of pattern isolation in Kant's formulation of the abstract categorical imperative ("Act only on that maxim through which you can at the same time will that it should become a universal law" [*Groundwork* 88]) and the related practical imperative ("Act in such a way that you always treat humanity, whether in your own person or in the person of any other, never simply as a means, but always at the same time as an end" [*Groundwork* 96]). There is a sort of patternlike beauty in seeing various small ethical preferences (not to kill, not to lie, and so on) as instances of a general principle that one should treat others as ends in themselves and never as mere means. But there is something more to Kant's insight than this. Specifically, if this were all that is at issue, the sense of the beauty (or sublimity) of the moral law would be contingent on accepting Kant's ethics.

Perhaps one thing that is going on here is captured in the sublime/beautiful distinction. In part, this is a distinction based on the component of aversive emotion in the experience. If there is a strong element of aversive emotion (e.g., fear and sorrow in tragic works), then we are disinclined to refer to the object as "beautiful" and to feel more inclined to use another term. But this may also point us toward another emotional component of aesthetic experience—what the Sanskrit writers referred to as wonder (Ingalls 16). It would seem that feelings of attachment and wonder or awe are partially mutually enhancing and partially mutually inhibitory. A fascination with the object figures in both, but awe may involve withdrawal tendencies, whereas attachment involves approach tendencies. Perhaps we would wish to say that the same general principles apply to both sorts of aesthetic experience, but that the proportion or relative intensity of attachment and/or awe guide the degree to which the experience is one of beauty or sublimity.

This still does not wholly explain the relation to morality. One issue is the degree to which our awe is for instances of following the moral law rather than for the moral law itself. For example, there is something sublime in a martyr who allows himself or herself to be killed in order to avoid violating a moral principle—as when Tom allows himself to be tortured to death rather than betray two escaped slaves in *Uncle Tom's Cabin*. Even Kant's abstract tragedy of the moral law is aesthetically resonant because it suggests particular cases of moral failure, including the reader's own, cases that are presumably stored in his or her emotional memory. They are also bound up with the reader's simulation of Kant's own despair, when he writes, "Morals began with the noblest attribute of human nature, the development and cultivation of which promised infinite uses, and it ended in—fanaticism or superstition" (*Critique of Practical Reason* 167).

FLA: Importantly, Patrick, you bring to the discussion the sense of degrees of presence of enhancement and inhibition in the encounter of the object. I would like to push a little more in the direction of aesthetic relations and subject-to-object relationality. In talking about aesthetic relations and responses, we might consider how this grows from a satisfaction—a pleasure—in accomplishment. It is very likely that this satisfaction became one of the central ingredients of what we later on established as an *aesthetic relation* with the objects that we build, construct, manufacture.

The moment that we humans became conscious of the fact that we derive not only a practical but an emotional satisfaction out of improving skills and innovating in manufacturing products such as the arrowhead, it is likely that we began to look at the products of our work with a certain amount of pride, to enter into a relationship of contemplation with the work—even if only fleetingly. These products were no longer just utilitarian. They were also experienced as satisfying, pleasurable—beautiful.

There are many other ingredients, including the neurobiological reactions of pleasure and even revulsion to certain kinds of food, which are also involved in the development of an aesthetic sense or an aesthetic relation with objects (objects of nature or the objects that we as humans manufacture). There are many factors involved in the development of what we call our aesthetic relationship with objects and with other humans. But there is nothing mysterious in the development of our aesthetic sense of things.

PCH: I of course agree that there is nothing mysterious in this—or, rather, that there is only something mysterious to the extent that we do not yet understand these matters. As to the origins of aesthetics, you are certainly right that pride has some place here. But pride presumably gives rise to the production of beautiful artifacts only because humans already had a predisposition to experience pleasure in pattern isolation. That enjoyment of pattern isolation is presumably what generated the decorative carvings initially. Or, rather, there could have been many sources for the initial production of what are in effect doodles. But as soon as the doodler began to take up a receptive attitude toward the doodles, he or she was likely to begin organizing them into patterns. A receptive attitude is simply looking at what one has done (or said) and considering it as an object, accessible to other people.[2] This receptive attitude does not create aesthetic feeling. However, it does allow aesthetic feeling to be provoked by encoding certain possible patterns. This is likely to lead the doodler back to the design, now with greater interest in developing patterns. It is, of course, possible that these designs had practical or magical associations as well. But, for our purposes, the aesthetic element is crucial (even if it was secondary or tertiary for the original artists).

As a doodler's skill grew, his or her work would likely have provoked positive responses from others in his or her group. This would, as you say, give rise to pride. That pride would give further impetus to the production of patterns. The point applies equally to sound patterns in poetry or to narrative structure in storytelling. But I would say that the pride merely adds another motivation to precedent aesthetic impulses.

FLA: Perhaps it is a sense of accomplishment or pride that leads to the contemplative instant—a moment that opens us to the aesthetic, artifactual emotion. Let me take a step back in the history of aesthetics, aesthetic theory, and its institutions to clarify. We can trace the origins of our awareness of what we call beauty or ugliness to the dawn of time—but it is not until about the seventeenth century in Europe that our aesthetic relationship with objects becomes a separate, independent autonomous field of action. This is when objects shed entirely all kinds of utilitarian or nonaesthetic features. Although we find this autonomy of aesthetics in an incipient way in ancient Greece and ancient Rome in sculpture, music (partially), drama, literature, the aesthetic domain, while it was recognized, was not yet institutionalized as a separate domain. We see the adding of an aesthetic element to Greek architecture and ceramics (plates and pots with drawings) that are effectively utilitarian objects.

While instances of an art-for-art's-sake sensibility and theory appeared sporadically in earlier periods, it was only in the eighteenth century that it was institutionalized. Art for art's sake explodes in the

---

2. In *On Interpretation*, I treat aesthetical intent in relation to this idea; in *Narrative Discourse*, I consider narratological implications of the idea.

eighteenth century. This is the moment when huge museums are built and when art not only becomes art for art's sake but becomes more and more universal. Countries like England and France begin to pillage artifacts from all over the world to bring to these huge museums objects that no longer fulfill any kind of utilitarian function or role. They play only the role of aesthetic artifact. Their sole role is to be contemplated and, being contemplated, to produce a sensation, a feeling of beauty.

In art for art's sake, we see the confirmation, the reaffirming of the aesthetic relationship with objects. The affirmation of this aesthetic relation that began in the Paleolithic makes this an autonomous domain. This is why museums are institutions specifically built to isolate objects, to render them devoid of any kind of utilitarian purpose. This allows for the *contemplation* of the objects as sources of beauty.

This history is perhaps our living proof of the fact that the aesthetic *is essentially a relation*—a relation between an object and a subject or of a subject to an object. Beauty is not contained in the object itself. It is not contained in the subject (human) itself. Beauty is contained only and exclusively in the relationship between object and the subject. A painting by Picasso tucked away in a vault not seen by anybody is not a work of art. It is not an aesthetic object. That object is only an aesthetic object when it is contemplated by humans. Or let us say you are standing in a museum and the lights go out. The paintings cease at that moment to be aesthetic objects.

What about literature, I ask? Perhaps it is easier to see the aesthetic relationship as a subject-to-object relationship in painting, but we see this also in novels. For the author's blueprints to be works of art, there has to be a co-construction (gap filling and the like)—an establishing of a subject-to-object aesthetic relationship.

PCH: I think we largely agree here, except that I would make a terminological distinction. As indicated earlier, I would say that an aesthetic *event* occurs only when there is a subjective aesthetic *experience* of a beautiful (or sublime) object. So, I would say that a painting in a vault may still be a beautiful or sublime object, but it is not part of an aesthetic event. This may seem to be a distinction without a difference. However, I think it does have some consequences. For one thing, in the co-creative account you present, it is not clear that we can actually assign properties to the work of art. Of course, even in my account, these need to be encodable properties. In other words, they need to be properties that could enter into an aesthetic event by affecting aesthetic experience. Nonetheless, there is, I believe, a difference between speaking of the patterning of colors or themes and variations, on the one hand, and the experience of attachment memories, on the other. In addition, this account allows at least some sorts of aesthetic criteria. For instance, it is often the case that Westerners new to Hindustani classical music hear the music as chaotic. In a co-creative view, it is difficult to say that this is mistaken. But, in my view, hearing, say, Rajan and Sajan Mishra's performance of Rāga Lalit as chaotic shows a lack of encoding skill.

As to the issue of history, there is certainly a change of some sort with the development of museums. But I do not think it is primarily a change in the degree to which works of art are contemplated. Rather, it seems that works of art have been contemplated since the beginning of time (as you indicate). Moreover, I do not believe that works of art cease having other functions—prominently religious functions. The issue is the degree to which works of art came to be located in particular economic systems and particular ideological systems. I would not interpret the colonial theft of art in aesthetic terms but in terms of market developments and ideological developments. For example, the influx of non-European art allowed Hegel to emplot the history of art in a way that placed "oriental" works at the earliest or most primitive stage, followed and superseded by Greek work, then Christian work. I do not by any means agree fully with Said's rather all-encompassing view of orientalism. However, on this topic, an analysis in terms of orientalism—modified by more standard ideological critique—seems to be on the right track.

HISTORY AND ART

FLA: I find it extremely productive to my own sharp, black-and-white contrastive thinking to engage with your degrees of complexity, Patrick. You also highlight the historical contexts in which this aesthetic history takes place. The violent colonial enterprise and its pilfering of objects the world over is just that. What I would like to further clarify is that our current understanding of an aesthetic relationship is only very recent and was only able to mature at a moment of human history when the artistic or aesthetic domain becomes itself an autonomous, independent domain: the moment when art shed all purpose other than the aesthetic finality. Only in this moment could we begin to talk about the general features, characteristics, of aesthetics.

It is not for nothing that it was in the eighteenth century that the philosopher Alexander Gottlieb Baumgarten wrote the first book with *Aesthetics* (1750) in the title—and this before Kant's *Critique of Judgement* (1790). In the early nineteenth century Hegel lectured on aesthetics in Heidelberg (1818) and in Berlin (the 1820s). A figure like Baumgarten could write a book on aesthetics only at the moment when human activity had created a separate domain for objects and their relationships with humans—what I have been calling the aesthetic relation.

We can understand, therefore, why there can be a separate study of music: we can study its formal architecture from a compositional point of view, we can study it as a biological phenomenon in terms of production and reception from the point of view of acoustics, and we can study it as a cultural phenomenon. The fields of ethnomusicology and sociomusicology more generally enable us to study music in the context of such and such political situation and even to talk about the politics of music, if so inclined, and we can also talk about the history of music.

We can bring many disciplines to bear on the study of music. But from the present angle of approach, what is important for us are the traits that music and the human relation to music have in common with, for instance, narrative fiction, cinema, and so on.

What they all have in common: they are all objects that have a very specific, special relationship with humans. They are products of human activity; but products that are identified and acquire a signification, a meaning (a sense) only in what I have been calling an aesthetic relation.

This is what I am identifying as a unified aesthetics, the idea that no matter how different the objects are, and no matter how different the specific conditions are in which they were manufactured, what they all have in common is that they stand in an aesthetic relation with humans. Humans and these objects have a specific relation that is different from all other relations, be they political, economic, juridical, or what have you.

What they all have in common, too, is that they can only be judged, appreciated, and understood *on their own terms* as aesthetic objects; that is, evaluated and understood in terms of universal categories such as the beautiful and the ugly or the tragic and the comic (as you point out in your work) and, I would add, the grotesque.

PCH: You are certainly right that the systematic study of aesthetics was likely to arise at a certain period. But I am not inclined to see this as primarily a function of developments regarding art. I am inclined to see it as a matter of developments regarding systematic study. There is something like a systematic study of art that seems to begin in the period you are discussing. But this is continuous with the systematic study of many other realms of experience. It is unsurprising that this would carry over to art. Moreover, the systematic study of, for example, literary art is not unique to the modern West. We find it, for example, over an extended period in the Sanskrit tradition.

As to functionality, many people do speak of a dissociation of art from function in early modern Europe. There is some element of truth in this—but it is, I believe, an overstatement. First, art in the past was not simply bound to functionality. It is true that the production of plastic art was often commissioned in functional contexts (e.g., for temples). But it seems clear that it was often dissociated from those contexts in experience. Certainly literature quickly sloughed off any ritual function. On the other side, it is not entirely clear that the functionality of art has been abandoned in the modern period. It is true that works of art are less likely to be commissioned by churches. However, at least in literary study, we highly value the political functionality of works. Indeed, it seems that our professional imperatives bear much more fully on such functionality than on aesthetic autonomy.

FLA: I see your point, Patrick. What I'm trying to formulate is a unified vision of aesthetics that seeks to understand the relationship between subjects (humans) and the objects we manufacture. Here the objects are not to be understood as instances or exempla (they are not in any way documents) to be judged and interpreted in nonaesthetic terms (politics, economics, or whatever) but are rather to be judged on their own aesthetic terms first and foremost.

PCH: Well, works can have many different purposes and many different values. One work may have political implications that counterbalance pernicious tendencies in a given society—for example, a work might effectively oppose the detention policies of the U.S. "War on Terror." Even if we found the work to be aesthetically lacking, I think both of us would find this valuable. Of course, in evaluating a work's politics, we are not evaluating it as an aesthetic object. That is important, since it does sometimes seem that our colleagues think good politics equals good aesthetics, which is as silly as thinking that good blood sugar levels equal good cholesterol levels. Indeed, this seems to me to suggest the degree to which we have not left behind the functional conception of art. Even professional literary critics seem often to feel that the main function of a work is in its contribution to some sort of worship.

FLA: This impulse to worship—to present hagiographies—certainly gets in the way of the work required for real criticism that sheds light on what works and what doesn't work according to reasoned analysis. I think it worth repeating my position: the aesthetic *is a relation*. It is not a property of an object, nor a property of the subject. It is the encounter between the object and the subject, giving rise to the particular reaction we call aesthetic. This aesthetic response involves a state of intense attraction we call beauty or a state of intense rejection we call ugliness. This means that beauty and ugliness are not in the objects themselves. They are in the relation between object and subject.

If there is an aesthetic module in the mind/brain, such a module would necessarily have to interface with the sensory motor system (mentioned in chapter 2) because this object/subject aesthetic interface requires the use of the senses—necessarily. There is only an aesthetic act or reaction in the face of an object—an object that has to be perceived by the senses to be evaluated. When the lights go off at the gallery, the relationship between me and the painting ceases to exist. My senses fail to perceive as such, and therefore there is no aesthetic. With language we have an optional interface with the sensory motor system: sometimes it interfaces, most of the time it does not. It is an option.

PCH: Maybe it is not worth quibbling over, but I am not sure I agree about sense being a necessary condition, even for an aesthetic event (in my terminology). I believe mere numerical ideas can be beautiful. On the other hand, I also believe that ideas always recruit images and I believe that emotional response is image-bound.[3] Given the close interrelations of number and space (see Dehaene), one might reasonably expect that the images associated with number would have an important spatial aspect. Thus it may be the case that an aesthetic event is impossible without some involvement of sensory cortex.

FLA: As you point out, the aesthetic relation (what you call the aesthetic event) requires the involvement of all the senses. That said, there are ways of using the senses *without using them*. If the relation with the object ceases— for one reason or another (lights going out while in a picture gallery)—

3. I discuss this in chapter 2 of *What Literature Teaches Us About Emotion*.

the senses cease operating and the aesthetic relation disappears. But there are other ways in which the senses can stop functioning within an aesthetic relation. Instead of being aesthetic objects, they may fall into the general category of simple objects—objects as such. For instance, I have a painting by Modigliani on my wall. It has been there for years. I no longer stop to *contemplate* the painting. I am no longer absorbed by the painting. We have already discussed this as the phenomenon of habituation. This, too, can potentially destroy the aesthetic relation. The painting on my wall becomes an object just like the chair or floor or baseboards in my house.

The way out of this habituation trap is what the Russian formalists (Shklovsky in particular) identified as the device or mechanism of *enstrangement*. It is through this mechanism—a device that reorients our sense of an object, that we make it new, that we "make a stone feel stony" (6)—that the picture hanging on the wall becomes new; you talk of this in terms of interest. The picture hasn't changed in content/substance, but my relation to it is renewed by this mechanism of *enstrangement*. This could be accomplished simply by moving the painting to another wall where I can spend time with it and contemplate it. Similarly, *enstrangement* is at work in Marcel Duchamp's readymades—his urinal housed in the museum, for instance.

Enstrangement can also be the result of *creating something new*—a different sort or kind of object that will in its turn create a new kind of relationship with the subject, a new aesthetic relation with the subject. The modernist expression comes readily to mind: "Make it new." Ceaseless search for the new (writing, music, painting, etc.) is precisely propelled by this aspiration of all artists to establish an aesthetic relationship with their audience, readers, listeners, and so on. As you state, we derive pleasure (reward) from this. Artists want to establish an aesthetic relationship, and in this ambition to create aesthetic objects and relations, they are constantly fighting against habituation.

García Márquez makes it new in *One Hundred Years of Solitude* (1967), and then the innovative device of magical realism becomes a mechanical application—a new form of habituation. The innovation that surprised and gave great satisfaction and pleasure to readers—that gave beauty—turned into a cookie cutter, a new form of habituation, from which boredom and the destruction of any aesthetic relation ensue. When I read Isabel Allende, I no longer feel like I am contemplating beauty. On the contrary, I feel bored. Roberto Bolaño puts it well: "Asked to choose between the frying pan and the fire, I choose Isabel Allende.

The glamour of her life as a South American in California, her imitations of García Márquez, her unquestionable courage, the way her writing ranges from the kitsch to the pathetic and reveals her as a kind of Latin American and politically correct version of the author of *The Valley of the Dolls*. [. . .] It won't live long, like many sick people, but for now it is alive. And there is always the possibility of a miracle. Who knows?" (110).

PCH: I understand what you are getting at here. Of course, I too have had the sorts of experiences you describe. I have three comments. First, I want to go back to a fairly simple but revealing model of emotion. According to this model, we have spontaneous emotional impulses that are largely subcortical. They may be triggered by external or internal events—perceptions, memories, and simulations. But we also have cortical monitoring and modulation of emotional impulses. Further, the monitoring affects the subcortical states through imaginative elaborations that provoke emotional responses, through recruitment of emotional memories, and so on.

Presumably, the same points apply to aesthetic response as to other sorts of emotion. If so, we would expect some part of our emotional response to be a function of prototype approximation, patterning, attachment, and so on. But we would also expect part of our emotional response to be governed by "judgment," to use the Kantian term (rather loosely), and thus by modulatory, cortical processes. (For a more faithful integration of Kantian aesthetics and neuroscience, see Linda Palmer.)

In this context, we might think a little more about the issue you raise regarding García Márquez and Allende. I do not want to take a stand on these particular writers. Rather, I wish to consider the general problem that you see as occurring in this case. Anyone who has spent time reading a wide range of literary works has almost certainly had the experience of enjoying a novel that used a particular technique (say magical realism), only to be disappointed in turning to a second novel using the same technique. However, I doubt that this is usually a case of habituation. In fact, it is often the precise opposite of habituation.

The sort of response you are describing frequently occurs in cases in which the technique is rarely used. In other words, it frequently occurs when the technique is obtrusive and attracts our self-conscious attention. In those cases, we may or may not begin to experience spontaneous aesthetic enjoyment. However, if we do, there is a good chance that we will modulate our initial response by a judgment about the originality of the work—or, more precisely, a judgment of the degree to which the work is derivative. Suppose that the difference in your response to García Márquez and to Allende is a matter of habituation. One result of this should be that you cannot reread Márquez, since the habituation should presumably inhibit your appreciation of his work a second time. Indeed, it should inhibit that appreciation more than it inhibits the appreciation of Allende. On the other hand, if your response is a matter of judgment, then the negative judgment need not apply to Márquez but only to Allende—particularly if it is a judgment of derivativeness.

An interesting result of this, which I discuss in *Joyce, Milton, and the Theory of Influence*, is that we are likely to become more accepting of a later work if many other people begin using the same technique. In other words, some literary technique may be pioneered by a particular writer. Suppose that writer has a single imitator, who takes up the technique. In that case, we are likely to judge the single imitator as derivative. But if the technique gives rise to an entire literary school or movement, then our criteria are likely to change. We may allow that there are several writers in the school who have merit. In a sense, we are able to respond spontaneously to individual authors using a technique once we have become habituated to the technique. Our inclination to modulate these responses diminishes with the decreased obtrusiveness of the technique.

On the other hand, even this is complicated. Specifically, you mention that, in your view, Allende is "mechanical." We do sometimes have the sense that an author is not "creating" but is simply following a formula. It is difficult to say precisely what constitutes this "mechanical" quality, though I think most dedicated readers have a sense of what it might be. Here are two possible components. The first is a relative paucity of simulation. Suppose the author is routinely following inferential principles in, say, understanding the minds of characters. We would expect this to produce characters who do not feel complex but rather appear "two dimensional." This is because the nuances of human attitudes and behavior are captured well in simulations. However, they are largely missing from general inferential principles. This is one reason why the portrayal of racial or other out-groups (e.g., blacks portrayed by a white author) are often unconvincing. Our tendency to think of out-groups as relatively uniform (Duckitt 81) is related to inhibited simulation and a greater inclination to infer mental states from generalizations. Parallel with this, when an author takes up a fashionable technique, it could very easily be that he or she ends up merely following the "rules" of the technique (e.g., magical realism) rather than successfully simulating the storyworld.

The idea of an author taking up a fashionable technique is related to the second likely fault here. That is, roughly, insincerity. If an author responds with sincere aesthetic enthusiasm to a new technique, that suggests that he or she has simulated a precursor's storyworlds and narrations in ways that are deeply congruent with his or her own engagements. This provides a partial basis for subsequent simulations. It also means that the new author should be able to judge the success of his or her own work. It will be successful if, on rereading, it provokes the same sort of aesthetic feeling as the original work. For example, if Allende responds with genuine appreciation to García Márquez, then that experience provides a sort of touchstone for her own self-evaluation. In contrast, suppose an author takes up a technique primarily because it has recently been successful in the book market, and thus is in demand by publishers. This author probably does not have the sorts of internal processes that would allow him or her to simulate a new storyworld in accordance with that technique. Thus he or she will be more or less forced to self-consciously craft some approximation to the technique. Moreover, this author will not be able to judge his or her own work intuitively, having no basis for comparison in a strong aesthetic response to the precursor.

FLA: Your thinking in complex conceptual overlaps and degrees raises important points that take into account factors such as marketplace and technique, Patrick. I will keep these in mind (albeit bracketed) as I continue to formulate a unified aesthetics. I should point out that there are several categories of objects with which we establish an aesthetic relation. We have, for instance, nature and the way we sensorially experience nature. I can walk outside my house and effect an aesthetic relation with the trees, birds, light, and so forth. I can also go outside my house and not be moved by any of this. There is also the phenomenon of habituation.

One of the strongest aesthetic relations we can have with nature is what the romantics called the sublime and to which you referred earlier: the deep admiration and awe and pleasure that the contemplation of nature can cause. The romantic would mention the sight of Mont Blanc, and there is the famous *Wanderer above the Sea of Fog* (1818) of Caspar David Friedrich. Even if the concept of the sublime were only used in the context of nature, in a unified aesthetics *there is* and *there can be* an aesthetic effect of that which is not created by man in the contemplation of something that is not man-made. We have an aesthetic relationship with all sorts of manufactured objects, such as arrowheads, which I already mentioned, and also the cave paintings in Altamira, Spain, or

Lascaux, France—all these can be the object of an aesthetic relation that goes beyond or that is parallel to the utility relation we have with them.

Then we have mass-produced industrial objects. In Italy we have companies like Alessi, which designs very aesthetically appealing espresso makers and teapots, Vespa, which makes appealing scooters, and, of course, Ferrari, which creates more out-of-budget aesthetic objects. In the United States, we have Apple computers—beautiful to the sight and delightful to the touch and intuitively utilitarian. They all serve utilitarian needs, but at the same time, they could be and are the object of aesthetic contemplation. They produce a sense of beauty, joy, wonder in us—especially, for me at least, the Ferrari 612 Scaglietti.

So we have two domains of objects: natural and man-made. There is, however, as I mentioned earlier, an additional domain that emerged in our earlier epochs, a domain of art-for-art's-sake objects, objects intended to be used specifically and exclusively for aesthetic contemplation. This art is not meant to serve any other purpose than to establish an aesthetic relation. It is only during this explosion that art becomes planetary art—there is no world literature, world music, and so on, before this time.

PCH: I think we may simply have to agree to disagree on the historical issue. Of course, there are changes in the degree to which different tendencies are prominent at different times. But I simply do not see Elizabethan drama (or eighteenth-century drama, or contemporary cinema) as more a pure aesthetics than, say, Sanskrit drama.

WHERE IS BEAUTY?

FLA: Fair enough, Patrick. Perhaps we can turn our discussion to a consideration of two concepts used to describe an aesthetic relationality: the content and the application of beauty and ugliness. These two concepts are linked to other typological or classificatory concepts that apply most clearly in literature but also, even if in a less clear way, in the visual arts and other arts. The classificatory concepts linked to the concepts of beauty and ugliness are essentially genres: comedy, tragedy, and the grotesque, and any number of combinations thereof.

Contemplating Théodore Géricault's *Raft of the Medusa* (1818–19) evokes a deep sense of tragedy—empathy for those on the raft and their ordeal—and at the same time the sense of the sublime, called up by the all-powerful action of the sea. This sublime of nature can sometimes be felt in its representation. Of course, the painting essentially depicts a tragedy—and that's the aesthetic feeling we get, the feeling of tragedy.

When a figure like Hieronymus Bosch comes along and paints the *Garden of Earthly Delight* in the late fifteenth to early sixteenth century, he makes the aesthetic relation with the object so new that it takes centuries for surrealists like Salvador Dalí, Leonora Carrington, and Remedios Varo to be able follow suit with the application of the grotesque in their paintings.

These typological concepts that are important in aesthetics and aesthetic studies appear in a weaker way in the other art forms as well. We find the grotesque very clearly in the foundational and scandalous (in the day) *Animal House* (1978) as well as in Robert Rodriguez's film mashups and Alan Ball's HBO show *True Blood* (2010–)—all of which attempt to break the unaesthetic relation of habituation with the object by turning this object into something new and full of life.

PCH: As you know, I take a different view here. I use "comedy" in a fairly ordinary way to refer to narratives in which the hero achieves his or her goal in the end. Conversely, I use "tragedy" to refer to narratives in which the hero's goal achievement is rendered impossible. The goals themselves are defined by emotion systems. Thus the goals can differ by emotion system. For instance, the goal may be romantic union or social authority. Moreover, our empathic emotional engagement with the hero seeking this goal is a different sort of emotional engagement than what we experience with beauty.

FLA: Indeed, your work on emotions and the expression of prototype narratives (the heroic, romantic, and sacrificial variants of tragicomedy) offers a serious and foundational research program. Nonetheless, I think it worth pursuing here a formulation of the concept of the grotesque.

The grotesque certainly allows for all kinds of exaggeration and for the carnivalesque (carnivalism) generally. The grotesque allows for the breach of all kinds of social rules and norms and values, such as in a Rodriguez film like *Machete* (2010) or *Planet Terror* (2007), within an aesthetic relation in which beauty is found in the transgression of norms and in the rehabitation of our perception of the object. Thanks to this, we have arts in many fields that dynamite the social consensus concerning, for instance, morals.

PCH: Certainly the grotesque can be combined with tragedy and comedy, as can mirth or jealousy or anything else. The issue is how to understand the grotesque in terms of neurologically grounded emotion systems.

FLA: Yes, let me see if I can work to this—or rather, through this. Each and every work of literature, film, comic book, and so on, is a construct (your simulation). To review several points: The building blocks of these constructs are the bits and pieces taken from the real world and organized and given shape with aesthetic intentions on the basis of the story and discourse. (See chapter 3.) In narrative fiction, the end result or product is a blueprint that requires the active contribution of the reader for it to become a fully existing work of art.

In all aesthetic relations, we have 1) a subject—the individual who creates the work of art; 2) the work of art itself that we can call the blueprint; and 3) the agent who completes the blueprint. An aesthetic goal is always in view in the mind of author and reader, and because of this, evaluation is always taking place.

All aesthetic relations necessitate an evaluation that establishes hierarchies of value, whether the object is a toilet in a museum or a novel by García Márquez. When reading a novel, while we fill in gaps in the blueprint, we are also constantly making value judgments. This can occur at the high level of analysis you bring to bear on authors such as Allende mentioned earlier.

An evaluation of the aesthetic relation does not happen out of time and place. When Joyce published *Dubliners* (1914) or *Portrait of the Artist as a Young Man* (1916) they were negatively reviewed and little read. Years later he found his readership—and this largely because of the controversy surrounding the censoring of *Ulysses* in the United States and Britain that stirred many a reader's curiosity. The same could be said of Nabokov. He was obscure even though he had written a lot; aficionados and some academics knew his work, but he did not become a household name till the scandal of *Lolita* (1955)—another book banned in the United States. (I'm offering a pattern, but your formulation, Patrick, could work just as well.)

We are talking about an aesthetic relation. It is not that *Lolita* possesses these virtues of beauty or any other virtues in and of itself. Those virtues only appear and only exist *in the reading of* Lolita—in the filling in the gaps of the blueprint and *in the global relation* between reader and work of art produced by the artist. The first generation of readers of *Lolita* and *Ulysses* did not ascribe to them aesthetic values such as beauty or well-made artifact or job well done, but the next generation did. Stendhal famously said something like "I am not writing for my contemporary human beings. I am writing for the future generations." Those today will not appreciate the work of art as those tomorrow.

There can be the phenomenon of habituation that we already discussed. A reader of detective novels may know the genre so well that he or she stops being surprised. They no longer find novelty in the use of different devices. Therefore, the reader's relationship with this kind of literature is no longer an aesthetic relation—a relation that may have existed before he or she accumulated a total understanding of this genre's generative devices. Yet this hasn't changed the object. What has changed is the reader's relationship to those kinds of novels; what has changed is the reader's aesthetic relation. And then there can be the phenomenon of blindness: the reader simply rejects the novel or poem or short story because it does not correspond to his or her expectations. In other cases, it might be that the reader does not have adequate knowledge of the literary devices used in the telling of a story and thus may reject the novel.

Evaluation is necessarily integral to the aesthetic relation, because the aesthetic experience is all about contemplation, shape, and meaning. The work of art is contemplated and experienced as an aesthetic object on account of its shape and of its meaning, the shape and the meaning conferred on it by the artist and perceived and contemplated by the viewer, listener, or reader, who is engaged in the act of contemplation by perceiving and appreciating shape and in their turn ascribing meaning to the object and its shape.

PCH: Interesting ideas. As you know, I do not view the work of art as a blueprint. It is true that no one will experience emotion in response to a work without his or her own mental activation of, for example, emotional memories. However, I would be loath to call those emotional memories part of the work. In architectural terms, the literary work is not the blueprint but the building. The building would be meaningless if no one ever used it. But I would hardly say that the building is just the blueprint for the building and the real building is what people do in it—or however this metaphor would work out.

As to the history of reception, many issues enter here. These range from professional jealousy to degree of interpretive and simulative effort to cognitive accessibility. Professional jealousy is the sort of thing you refer to with Woolf, which is possible (I do not know much about Woolf's personal attitudes). Lack of cognitive accessibility is the excessive distance referred to by Jauss. For example, when interior monologue was introduced, readers did not have the processing techniques readily available to them that would facilitate understanding. By "degree of interpretive and simulative effort," I am referring to the fact that any complex piece of writing requires work. If we know nothing about the author, we are unlikely to want to invest much time in effortful interpretation and simulation. This is in part for the simple reason that we do not feel confident that the work will result in an adequate reward. Once the work is established, however, we are more likely to risk the effort. Indeed, in that case, the precise opposite may occur. We may overvalue the work, deciding that it was worth the effort, despite rather feeble results. The point applies not only to fiction but to literary theory.

FLA: Perhaps it is this lack of a work ethic, say, in reading something like *Ulysses* that prevents (blinds?) them from entering into an aesthetic relation with it? No matter how much I explain the aesthetic pleasure of *Ulysses*, my students just do not get it. They do not get why I am giggling when Bloom is in the outhouse, farting and pooping while contemplating life; they do not see the aesthetic virtuosity of the moment.

    I wonder, too, if the rejection of the aesthetic relation occurs most frequently with respect to works we classify under the grotesque, its main ingredients being the carnivalesque, an irreverent attitude, and the use of the ugly and the repulsive. Comedy and tragedy do not deliberately use the ugly and the repulsive as the aesthetic means to attain beauty in a strong aesthetic relationship with readers, but in the genre of the grotesque, this is very common. This partly explains my attraction to Tex Avery cartoons and comic-book films such as those of Robert Rodriguez.

PCH: The scene in the outhouse does indeed have its own form of catharsis.

FLA: Now, perhaps, I can take a stab at your earlier injunction: to consider where the neurobiological fits into all this. Suzanne Nalbantian speaks of authors/artists as living people with mind/brains; she does not ascribe a neurobiology to a character in a novel or in a painting. She keeps these levels very distinct. She writes, "The integrative process of art is characterized by 'parsimony'—or the distillation of complex distributions of data to simple aesthetic patterns. The higher-level synthesis, involving mental selectionism, links reason and pleasure or sensation in the operations of the brain" (357). This describes well what happens when I have my own particular experience that belongs only to me in a specific time and place and how this experience can only be communicated to you when the experiences are brought to a level of a gestalt. Moreover, what I communicate is not the "complex distributions of data" but is a *reprocessing* of my experience. I distill and bring to the foreground the essential aspects of the experience. When I share with you my sense of a Robert Rodriguez film or a novel by Faulkner, I do not try to reproduce my idiosyncratic experience of every second of the movie or of the novel. I reproduce a distillation of the plot and of the aesthetic goals and means behind the creation of the work and its recreation in my mind.

To put it otherwise, to make the aesthetic experience manageable and to go beyond idiosyncratic experience, we have to use the mental faculties that allow us to transform this idiosyncratic experience into a series of aesthetic patterns or gestalts. To do this, I might use concepts such as the aesthetic concept of the tragic, comic, or, in the case of Rodriguez, the grotesque, for instance. Such a concept gives me a tool to analyze the relation I have with the film, novel, comic book, and so on.

PCH: It seems we diverge in our aesthetic judgments in some of these cases. As you know, when I speak of "universality," I am not positing a universality of response. I am referring only to a universality of structure and processing.

FLA: Beauty *is not in the eye of the beholder* (subjectivist and idealist). And beauty *is not inherent in the object* (materialist). Beauty is in the relation between subject and object.

PCH: Again, this is primarily a matter of how one is using such terms as "beauty."

FLA: Together with common everyday human thought and action, deep archeological and other evidence indicates the presence of dualism as an ordinary, normal approach to the world as a whole. The cave paintings in Altamira and Lascaux and anthropological understanding of customs such as the burial of the dead, for instance, corroborate the presence of dualistic thought long into our past.

We can say that since the dawn of time, Homo sapiens has interpreted the world in dualistic terms. This approach and worldview became systematic through magic, religion, and philosophy—and of course, such a predominant worldview has to a certain extent permeated science.

It is not unlikely that dualism has a basis in certain material functions of the mind/brain. For instance, the moment we are capable of understanding the functioning and application of the principle of causality in the natural world, we are also probably able to apply the principle of counterfactual thinking. (We have talked at length about this already in chapter 1.) Thus, in seeing life end, we can also imagine life continuing beyond life's natural state. That is, I can conceive of life continuing beyond nature. This is no doubt the basis of dualism: in magic or magical thought, in theology, and very influentially in philosophy beginning with Plato.

Essentially, the different theories that were developed from Aristotle on concerning the mind and its workings were essentially dualistic theories of mind. Only since the Renaissance and then the Enlightenment has dualism been seriously challenged. In the Renaissance this challenge ran parallel to the development of science (physics and biology), and in the Enlightenment it ran parallel in the struggle against religion and the ambition to apply the scientific method to all domains of life.

Since the Renaissance, the forms of dualism most strongly dependent on magic, theology, and anti-scientific outlooks of philosophy have lost ground. Just as we can say that in the last twenty-five years we have learned more about the functioning of the brain than we knew in all our prior history, we can also say that in this same period, science in general has achieved important victories over dualism and dualistic approaches to nature, including humans and their socio-neurobiological nature.

This does not mean at all that dualism has been entirely surmounted. Not at all. The so-called mind/body problem, as it is more commonly known, is still discussed, examined, and tested in many serious ways without as yet achieving definitively satisfactory conclusions.

The unitary approach to aesthetics that I propose here and my ambition to draw boundaries around a unified aesthetics is of course an aim and a methodology that inscribe themselves entirely within this unitary view of nature as a whole. I am not going to be a dualist in biology and a unitarian, say, in aesthetics.

PCH: You know my views on dualism. Put simply, the self is the (subjective) condition for all discourse about causality. As such, it cannot be part of that material causality. In quantum mechanical terms, it is always and necessarily outside the system being explained.

FLA: Let me offer a final—but not last by any means—word on aesthetics. From the unitary point of view, it is useful to refer to some scientific findings in the field of neurobiology that concern the so-called plastic arts, music, and narrative fiction in its very diverse media and guises.

This research that connects aesthetics and the socio-neurobiological sciences is quite incipient. The findings are quite tentative. Still, there are a certain number of observations, hypotheses, and theories that have a more and more solid grounding.

In philosophy, aesthetic theories—partial, limited formulations of which we find in Plato and Aristotle—are essentially dualistic, notwithstanding Aristotle's naturalistic realism. According to Plato, all aesthetic objects as well as all aesthetic concepts find their true reality or their truest nature not in this world but in the world identified as the *topos uranus*, where all that pertains to this world has its ideal counterpart. Of course, Aristotle was a materialist and did not go the idealistic route. Yet, he saw all artistic creation as *an imitation*—as a mimesis of the real world. Like its many other variants through history, this view posits that we have reflectors or mirrors that simulate the real world, so the world is in effect split in two. It is more naturalistic in explanation than Plato's theory, but it is still dualistic.

In a unified aesthetics, from the point of view of the aesthetic relation in which the aesthetic is not in the object nor in the subject but in the relation between the two, I identify the steps we follow in *any creative activity*. Beauty is not in the object, it is in the relation between the subject and object. Our distant ancestor who chips away at the rock to make an arrowhead experiences satisfaction in the making and improving of the arrowhead. This same ancestor contemplates the object and considers its beauty. In an act of contemplation, this ancestor would consider how well the arrowhead was made. In so doing, this ancestor modifies his or her relationship to the arrowhead because he or she considers it, even if for a moment, as an aesthetic object.

With a poem, short story, novel, sculpture, or whatever, we do not create an imitation of reality but bring a new reality to the world. We bring to the world a new approach, a new relation to the new object made in the world. An object made to kill becomes an object of contemplation, even if fleetingly. As society develops, some become more and more specialized in the act of contemplation, in making objects with an aesthetic intention and contemplation at a distance.

I already mentioned Shklovsky's formulation of *enstrangement*—his making of "a stone feel stony" (6)—but let us not forget Bertolt Brecht's small treatise on what he called his "epic theater." The analysis he makes of this is precisely a discussion of the *distance between the object and the subject* required to establish an aesthetic relation: if the relation is too proximate or too close, then no aesthetic relation is established. I just watched Rodriguez's *Spy Kids 3* (2003), a film in which he tries to make cinema work like video games and to force it into the aesthetic of video games. He thereby establishes a small distance (and at times no distance) between the one medium and the other and so ends up destroying the aesthetic relation. As an aesthetic object, the film is deeply flawed. I often visit the Wexner museum here at The Ohio State University. Occasionally, I wonder what has happened to the paintings. They hang on the white walls, but the walls are so big they dwarf—even nearly make disappear—the art. This is why museography has become a key field in the adequate presentation of aesthetic objects. A bad museographer can doom a work of art—even an artist and his or her career.

CHAPTER 5

Situating History, Culture, Politics, and Ethics in Literary Studies

PATRICK COLM HOGAN: I thought we might begin our final conversation with the issue of interpretation. In literary study, the idea of interpretation seems to have gotten rather badly convoluted in recent decades. It sometimes seems that the purpose of literary interpretation is to use whatever rhetorical technique will persuade a reader that a text should be read in what initially seems an entirely implausible way. I am exaggerating some, I know, but only some. There are presumably economic reasons for this. A new reading is, in effect, a "revolutionary new product" that obsolesces whatever brand you formerly had in your home.

FREDERICK LUIS ALDAMA: I must forcefully state that I wholly agree with you here. In fact, among scholars in our field there is so much confusion that often the object itself of interpretation is not clearly identified. Often, too, we do not even know what the interpretive activity is addressing. We come across notions such as "discourse" (in the Foucauldian or the Chatmanian sense?) or "text" (in the Derridean or the Jakobsonian sense?) and a whole multitude of other ideas that are supposed to be concepts used in enriching our understanding of the object of the interpretation. Most scholars working in the field of literature complicate enormously what they are doing and what interpretation means because they *do not* clearly identify the field the instrument of interpretation is applied to. To make any progress in interpretative work, a lot of branches need to be cut and discarded.

ON INTERPRETATION

PCH: Related to this is a strange belief that seems very common in literary discussions. This is the view that saying interpretations are equally valid is somehow "democratic." I take it that a system is democratic if it allows everyone to voice and make a case for their views. A system is not made democratic by claiming that all views are right.

FLA: Unfortunately, Patrick, scholars and even universities as a whole are sometimes more interested in making money than in doing empirical research and seeking knowledge. This is the basis for the academic star system and the investing of seeming importance in what amounts to passing fads.

It is the star system that often allows for this lack of responsibility toward the making of actual knowledge. It is this system that fosters the growing of the anything-goes approach with respect to interpretation. This has given the humanities—and in particular, the study of literature—a bad reputation in the eyes of colleagues in other departments. There is much hard work required in generative research programs in the sciences and humanities—those that may lead to a unified knowledge—about us, the things we make, and the world we inhabit.

Since scholars of literature have been given a free pass with this anything-goes approach, those outside of the academy feel as if they too can give as adequate an interpretation as those who study literature professionally. This is possible because the term "interpretation" is used in an extremely vague way. Since all aesthetic acts and relations and experience are also experiences in which shape and meaning are present as necessary ingredients, then meaning always implies interpretation. We have to understand what a phrase means to appreciate its difference from another phrase. Since this is implicit in all our activities—not just in aesthetic production and reception—that means there are grounds for the idea that anybody out there can *do* literary interpretation and for dismissing those who get paid to do this.

When the professional study of literature becomes more like the interpretation that happens while reading on the beach (for different reasons, including foremost that one no longer cares about rigorous empirical research and knowledge seeking), then the line between professional study and idiosyncratic reading blurs. This said, I remind myself of a moment earlier in our conversation when I mention that all are welcome to derive their pleasure from wherever they would like in their approach, method, and interest.

PCH: We'll return to democracy and equality below. For the moment, we might consider the usual goals and conditions of interpretation in ordinary life. When you say something to me, I need to understand it. Understanding your utterance means comprehending two things—its meaning and its purpose. So, suppose I ask when you will be able to read through our first version of this dialogue and you say, "Unfortunately, the book on Rodriguez is due at the end of this month." In order to understand this utterance, I need to know, for example, what the book on Rodriguez is. Among other things, I have to know that you are writing a book on a particular filmmaker. It would obviously be different if you had borrowed a book on the filmmaker and simply had to return it to the library before the end of the month. I also need to understand why you are telling me this—presumably, it is to communicate that there will be a delay of a few weeks before you can get to this manuscript. These components of understanding (meanings and purposes) constitute the goals of interpretation.

On the other hand, we do not engage in explicit articulation of all meanings and all purposes. For example, if the preceding were a real case, I would probably not engage in explicit interpretation at all. Your meaning and purpose would not require this sort of explication. However, suppose you send the manuscript to me and ask when I'll be able to get to it and I respond, "Unfortunately, the Iron Man Marathon is at the end of the month." You probably will not process this automatically. This may give rise to a quandary that leads to explicit interpretation. First, perhaps you are not familiar with Iron Man competitions. Second, at least for a moment, you may wonder if I am actually in training for such a competition. Of course, the moment my frail, waifish figure appears before your mind's eye, you will realize that I cannot possibly be preparing for anything so physically robust that it would contain the phrase "Iron Man" in it. But then the question may arise as to just what my purpose is in saying this. Am I indirectly chastising you for taking a few weeks with the manuscript? That seems uncharacteristically mean for such a characteristically nice fellow. Perhaps I am making some sort of disappointing attempt at humor?

FLA: As you so nicely lay out, the two components of understanding are meaning and purpose. I would add to this binary a third component. Understanding also has shape as a component. (I mention this in our chapter on aesthetics.) As your example shows, the end could be me or you saying, "I am just joking." To perceive the humor, the sentence or series of sentences—the utterance—has to have a humorous shape. It has to have a material, discernible, specific shape that distinguishes the humorous from the serious purpose or intention. What makes some explaining relevant and even necessary? If you tell me "There is an Iron Man Marathon this week," eventually I will understand the meaning and the purpose of this utterance if its relevance is made clear. The utterance by itself is not enough for me to understand its meaning and purpose. I do not know what an Iron Man Marathon is. I do not know the referent of this sentence. The purpose of communicating this to me is also obscure. To identify a correct meaning and purpose I need additional information; otherwise I will interpret it by assigning a degree of relevance that gives it a meaning that may or may not coincide with the intent of your communication.

If you asked "What Rodriguez book?" in reply to an email I sent, and I wrote back "And your mother, too," you might interpret this in many different ways—you might conclude that either I have lost my marbles or maybe even that I mean by it a very insulting remark to the effect of "Take a hike, pal." A neutral phrase like "And your mother, too" can be perceived differently if I ascribe to it the meaning "take a hike, pal." There are many instances in which we use euphemisms or phrases shaped as neutral phrases that are nevertheless deeply insulting. Like the game of *albures* in Mexico there is the African American tradition of "the dozens" (spectacularly done in Ellison's 1952 published novel, *Invisible Man*, for instance). The winner of the dozens is the one who can hurl the most insulting phrase in an aesthetically appealing way. The winner of *los albures* is the one who can deploy the most coarse insults within the structure of a rhyme. So the presence of shape also plays an important role.

PCH: Your metaphor of shape is interesting. As you know, my view is that a humorous event has to do with processes that are characteristic of childhood.[1] In the case of meaning, this would involve the overgeneration of

---

1. The full account is, of course, more complex than this suggests; see chapter 5 of Hogan, *What Literature*.

contextually nonrelevant meanings, a right-hemisphere-based activity, related to children's language processing.[2] In addition to such overgeneration, mirth is enhanced (or perhaps simply not inhibited) when the resulting ambiguity is interpretable for each meaning. Thus a successful pun should allow for some discourse coherence in both senses. Take the joking title and author *Life Under the Grandstands*, by Seymour Butts. It does not really work with the authors Seymour Foot or Seymour Brain, though those work as jokes with the different titles *Life as a Shoe Salesman* and *Life as a Neurosurgeon*. This enhanced discourse coherence or increased semantic relevance is perhaps something gestured toward by your metaphor of shape.

So, if we now turn to literature, we may draw some broad connections. First, when we interpret meaning, we might say that our aim is to construct the storyworld—the characters, events, motives, causal sequences, and so on. Obviously, meaning has to do with such matters as lexical items and syntax. But we might say that our most encompassing concern about meaning is getting the storyworld right. This "rightness," I should note, involves getting the ambiguities right as well. For example, if a detective novel does not make clear who the murderer is, then the uncertainty is part of the storyworld. Similarly, if the novel strongly suggests that Smith is the culprit but leaves hints that it might be Jones, then getting the storyworld right entails getting this ambiguity—probably Smith, but maybe Jones.

FLA: Yes, getting it right, as you put it, involves lexical and syntactic shaping as well as other ingredients ascribed to the work in its parts and as a whole. So what do we mean when we talk about meaning? Is the global meaning of the short story or the novel a total sum of the meanings established and conveyed in each of the phrases or sentences that form its linguistic building blocks? Is it in the identification of theme or character or, as you put it, in the making clear who the murderer is, so to speak? Where is it located?

Meaning is a very elusive concept. Every philosopher has his or her pet theory of "meaning." After World War II, the most prestigious theories of meaning were formulated by the logical positivists, Wittgenstein, Searle, and Derrida. In *Why the Humanities Matter* I go more deeply into a discussion of their theories to determine which are true to the facts and

---

2. See Hogan, *What Literature* 147–48 and citations, particularly Chiarello and Beeman 248, Beeman 272, Chiarello 145, and Kane 41 and 43.

more scientifically acceptable. Here, however, I would like to continue to follow Chomsky's formulation. For many years, Chomsky's position was that it was located in syntax. Therefore, it was impossible to build a theory or science of semantics. That the task at hand was to build a scientific theory of syntax. Once this work was accomplished, we would have a proper understanding of meaning. As you know, he left this position behind and developed the theory of minimalism to account for meaning.

I'll try to be as specific and clear as I can in the use and elucidation of the concept of meaning. Many have linked meaning to culture. I link meaning to truth. We need to know what the *meaning* of meaning is in order to submit the concept to empirical verification. If we do not get this right, then *anything* is meaning. We need to keep the discussion at a high level of generality but *as precise as possible*. Our conversation certainly aims to achieve this.

NARRATIVE PURPOSES

PCH: A story may have many sorts of purposes (the second component of understanding). But two general types of purpose recur extensively. As noted earlier, authors most often aim to produce an emotional effect and/or to convey some thematic concern. We may distinguish three sorts of emotional effects, related to different aspects of narrative. The first is story emotion. This is our emotional response to characters and events in the storyworld. It is the topic of most of my own work[3]—but I am far from alone in this. This topic has concerned literary theorists for thousands of years, as Aristotle's discussion of fear and pity and the ancient Indic writers' examination of *rasa*, or empathic and aesthetic emotion, suggests.

The second sort of emotion is plot interest. "Plot" (as you obviously know) is the presentation of story information—its selection, ordering, and construal. This discourse-based manipulation of story information regulates what we know about the storyworld and when we know it. Such manipulation of course has effects on our story emotions. However, it also trains our attention on clues regarding information that is missing—thus our feeling of suspense about what is to come (as when the innocent little girl enters a room where we have just seen the villain plant a bomb) or our feeling of curiosity about what has already happened (as when the beloved crusty old ruffian turns up again after many years

---

3. For example, it is central to *The Mind*, *Affective*, and *What Literature*.

but now with a peg leg and a marble eye). (See Meir Sternberg for a technical account of "suspense" and "curiosity." Writing in a cognitive context, Ed Tan has also stressed the development of interest.)

The third sort of emotion is sometimes called "artifactual" (see Tan 65, 82; his category of "fiction emotions" is parallel to what I am calling "story emotions"). This is the aesthetic feeling that we experience in response to the work as a created object. This clearly includes aspects of story and discourse. It also prominently includes aspects of style.

FLA: You nicely keep separate the author-reader emotion and the artifactual, or, as I would call it, the aesthetic relation (which is constituted by the aim of the artist, the aesthetic product, and aesthetic reception). I think of an author like Dostoyevsky, who clearly wanted to convey a worldview and within this worldview a political position concerning the Russia of his day. But if his purpose had only been to convey this content or message and to generate emotions in the reader, he could have written pamphlets, using the most effective tools furnished by the discipline of rhetoric. The many sentences that fill out the *Communist Manifesto* (1848) are beautifully composed and display a very effective use of rhetoric. In Lincoln's speech at Gettysburg we see the same careful rhetorical crafting of something that is not literature.

Dostoevsky, Flaubert, Faulkner, Woolf, Hemingway, and Stein all had centrally in mind their artistic purpose. What they wanted to do was create an artistic object, an object that would be aesthetically appealing. If their purpose had only been to generate emotions and convey themes, they would have worked along a different register, writing a rhetorically effective manifesto, speech, and so forth.

Of course, the vast majority of novels out there today represent a very diminished "will to style," but any author worth her salt intends to create an aesthetic object with an aesthetic purpose. The author develops her thinking and carries out her activity in such a way that the aesthetic object will have an aesthetic purpose. As you mention, this is in the style.

PCH: In speaking of emotional purposes, I did have in mind a reader's relation to the work itself—the storyworld, the discourse, the style. But both purposes of an utterance also concern a reader's relation to the world outside the work, the world of real life. Authors wish to affect the reader's understanding of and emotional response to that real world. Most often, that understanding and response concern either ethics or politics. We may refer to the particular understandings of and emotional responses to the world as "themes." Although not referred to in these terms, political themes have been at the forefront of professional interpretation in recent years.

FLA: Yes they have—and as you've pointed out so judiciously earlier, this can lead to opening and closings in our thinking, and has. I ask: what do we learn from narrative fiction in its most expansive sense—novels and short stories but also comic books, cartoons, films of all sorts, and much more? Do we learn anything about actual reality in the first few years of the twenty-first century from Rodriguez's *Spy Kids* (2001)? Or, under the influence of the film, are our emotions set in motion to defend the cause of the exploitation of Fooglies (odd-looking transmogrified humans who speak backward)? I am talking about a successful aesthetic object in this film. We could also ask, what is the political purpose of the horror film *Saw I* (2004)? A few survive out of a group of young people abducted and slaughtered, and they manage to kill the sadistic killers. What does this film tell me about Iraq or Afghanistan, where there is also a lot of violence and torture?

Put simply, and contrary to much literary criticism, I do not go to the cinema to learn history, sociology, politics, and so on. Most so-called period dramas, television shows, and films are very far from historical reality. What does the historically set *Generation Kill* (2008) that follows a marine Alpha team during the invasion of Iraq have to do *with the invasion of Iraq*? What is compelling is its carefully orchestrated banality as a fictional blueprint. We do not confuse reality for fiction where fiction takes the upper hand.

THE FUNCTIONS OF CRITICISM

PCH: Of course, you are completely right that we do not (usually) confuse reality with fiction as we are reading or watching the fiction. However, in general, we have very poor "source memory" (see Schacter 114–29) and thus very poor memory for the origin of memories—whether in fiction or fact. Daniel Schacter discusses research showing that narratives explicitly identified as false are often remembered as part of an actual event (115). Following D. T. Gilbert, he explains that "it requires a good deal of effort . . . to muster the critical faculties to 'unbelieve' new information" (117). Moreover, there is evidence that fictions alter even well-established beliefs in retrospect. Prentice and Gerrig discuss how people share a "tendency to allow any information, reliable or unreliable, to gain entry into their store of knowledge and to influence their beliefs about the world" (530). In keeping with this, "the mass-communication literature contains many findings of attitude change resulting from exposure to fiction. For example, numerous studies have shown that media portrayals of sex-role and ethnic stereotypes affect children's attitudes toward women and members of minority groups." Indeed, "a single exposure to a feature film produced significant attitude change (e.g., viewing *The Birth of a Nation* affected attitudes toward blacks)" (531). Thus it seems fairly clear that fiction does affect our understanding of the world. It also affects our emotional/motivational orientations.

Returning to interpretation, we might say that interpretation of narrative should have as its basic function the clarification of the storyworld (within the scope of its ambiguity), the elaboration of the emotional purposes of the work (including story emotion, plot interest, and artifact emotion), and the explication of the moral and political implications of the work (as these extend to the understanding and evaluation of the real world). These functions of interpretation arise in particular conditions, when we find our understanding of a work's meaning or purposes in some degree baffled.

But when is our understanding baffled? When do we have a sense that there is something that requires interpretation? This may occur spontaneously, of course. But it may also occur when we reflect on a work.

This is a point at which criticism enters. Drawing on cognitive science, we may say that *criticism* occurs with any practice that leads us to encode aspects of a literary work (film, work of art, or whatever) that we would not have encoded spontaneously. As I put it in *Cognitive Science, Literature, and the Arts*, encoding involves selection, segmentation, and structuration. In other words, it involves picking out aspects of an object (we can never experience all aspects of a work), organizing them into units, and integrating them into broader patterns. Thus criticism leads us to notice certain elements and relations in a work that we would not have noticed otherwise. As such, criticism is likely to give rise to interpretive questions.

For example, many ordinary readers of *Hamlet* may not notice that Hamlet suggests that he is going to murder Rosencrantz and Guildenstern well before he claims to have learned that he himself is to be killed in England (see 3.4.207–9 and 5.2). Moreover, it is not clear that Rosencrantz and Guildenstern were aware of Hamlet's fate. Thus it is unclear that they were guilty of anything that might justify their murder (even leaving aside procedural issues, such as a trial with right to respond). Finally, ordinary readers may hardly be aware that their death is announced only shortly after Horatio has expressed the hope that Hamlet's spirit will go to heaven. This conjunction, combined with the suggestions of premeditated murder, makes it at least possible that Hamlet's death is a death of his soul as well—it raises the possibility that his soul is guided not by angels (as Horatio hopes) but by demons. We may say that the text invites us to interpret these points. But we will interpret them only once we have encoded them, and that may occur only following criticism.

FLA: I see that what you call encoding, I call shaping. As I have proposed, I consider that all things cultural are prey for interpretation. That said, I know that we are focusing on literature here. And here you nicely identify how the interpretation can be focused at the level of the story and the domain of our bafflement. But before we even begin engaging in interpretation we have to decide what the domain is in which we are going to apply our interpretive faculties. That is, I suggest we study first and foremost the shape-giving elements—the generative operator of discourse. A work of art is not defined by its content. A work is circumscribed by the way the artistic or formal devices, as you wish, are brought to bear on the subject matter.

When I analyze a ballet, for instance, I ask what the main components of my analysis are. On the one hand, there is music, but this is a specific arrangement of sounds, so I study the specific shape given to one sound after another after another. That is, I study the arrangement of the composition. I study the body movement of the dancers—head, torso, legs, arms, and so on. That is, I analyze the organized movement of the dancers. So my main focus of analysis is the formal movement: how one posture follows another, and so on. I select, segment, structure, to use your words.

I do the same with all the other arts. For each novel that has been in one way or another intellectually, emotionally, aesthetically satisfying, what remains in my mind are the devices more than the details concerning the story. This is so because art is essentially a conferring of shape to matter. In the case of narrative fiction, it is the conferring of shape to subject matter.

In a theory of narrative fiction, when we consider fiction as a work of art, the starting point is the shape-giving tools. What are the shape-conferring instruments used and applied to a subject matter?

We need to keep in mind that the shape-giving tools do not give shape to a separate (or preexisting) subject matter. Rather, the matter becomes an inseparable component to the shape-giving instruments. That is, matter and form make for an inseparable unit. We can separate them for pedagogical reasons, but as a work of art they constitute an organic whole. *The Sound and the Fury* shows this clearly. Faulkner has an image (the subject matter), and as he gives shape to this image, he is simultaneously creating and multiplying his subject matter. This image (soiled underpants) needs to be placed within the story: Who is seeing the soiled underpants? It is not the girl, so Faulkner invents the character of Benjy. And to flesh out Benjy, other characters are invented, including brothers Quentin and Jason and a sister, Caddy. All of this is given even greater shape with Faulkner's use of focalization and other devices.

The subject matter is a systematic expansion, and each element (item of the story) in the expansion that is added coheres within an artistic whole. So each decision Faulkner makes in the additions—the point of view that he adopts and the other formal mechanisms that he uses—he wants, to form a coherent, structured whole. Faulkner also has to take into account the limitations of the reader (memory, primarily) in order for the three parts of the novel to form an organic whole—and for it not to appear as the mechanical addition of three hermetically sealed sections. This is a decision pertaining to form, and this has storytelling implications. That is, it has thematic implications. With the shift from first to third person, there is the possibility of investing the same characters of the first section with greater mental complexity. Formal procedures lead to the story matter, and when the story matter is more complicated, a need to look for other means for giving shape to the story matter arises.

All the formal decisions have material implications. The nature of the story changes when the author changes the formal methods used for the storytelling. This is further proof of the unity of form and matter, matter and form. Bearing this in mind guards against one slipping into the trap of considering the work of art as a document, an instantiation of a political or moral position—as if we could separate form from matter and speak of the novel (or whatever narrative fiction we happen to be analyzing) as a pamphlet.

In the Renaissance, painters invented perspective. This formal device was considered an important progress in the general domain of painting because, according to the aesthetics of the moment, perspective made painted reality *more real*. It increased the realism and therefore gave realism as an aesthetic category a central place in painting. Realism using perspective became the expected way to paint. Then, at the beginning of the twentieth century, Braque and Picasso did away with realism, using a new device called cubism that made visible what is invisible in realism, such as the showing of fronts and backs of heads simultaneously. They created a different form of realism that disconcerted viewers; viewers considered it ugly, flat, meaningless, and concluded that the artists had no idea what perspective was. In other words, the artist invents the device of the perspective, and this has consequences for the subject matter and also for what can be shown. The choice of devices in cubism, however, allows for different possibilities. The artist is no longer held accountable to the aesthetic rules of realism based on perspectivism and can instead follow a different aesthetic procedure and subject matter and therefore a completely different rule.

Here again we have a unity of form and matter. In accomplished fiction, we have this unity. We can see how in this unity of matter and form, form can *secrete subject matter*. It is as if the artist were a special kind of spider, a spider who wants to create something, and every time the spider makes a formal move, the result is a new shape and new content in the form of the web. What we have in the making of art is form that secretes matter—and matter, in its turn, directly influences the form.

Whether we focus on painting or music or ballet or literature or comic books or film, among many other art forms, they are all activities that require a myriad of decisions on the part of the creators. And when an artist makes decisions concerning form this necessarily impacts the subject matter.

PCH: Criticism, as defined above, may of course apply to any of the arts you mention. Within literary study, it may apply to any "level" of interpretation—storyworld, story emotion, plot interest, artifact emotion, ethical or political theme. Clearly, we want to spend at least some of this chapter on ethical and political themes. However, before turning to those, I would like to consider the application of social criticism to the storyworld. Specifically, it is clear that the dominant trend in current literary criticism is historicist and culturalist. In other words, when critics approach a narrative, they are concerned with altering our encoding of that work by reference to historically and culturally particular ideas, events, and practices. This is, of course, very important. As you know, a great deal of my own work on postcolonial literature and film involves relocating works in their cultural contexts.[4] To take a very simple example, a non-Indian viewer will probably misinterpret the storyworld of a film, perhaps significantly, if he or she sees a young woman with red in the parting of her hair. Indeed, such a viewer may not encode this at all (i.e., again, select and integrate the information with cognitive processes—a different idea than your "shaping," I believe). In any case, he or she is unlikely to recognize that this coloring indicates that the woman is married. For sometimes very simple reasons (such as this), and sometimes for much more complex reasons, cultural context is imperative for the criticism and interpretation of narratives.

FLA: Once again, I think it is important to be clear about the level of abstraction and therefore generality and even universality in which we are situating our analysis. Yes, interpretation can play an important function, but it is not working at the same level of abstraction as a scientific theory of narrative. Theoretical physics belongs to a level of analysis—and therefore generality and universality—that is quite higher than so-called applied physics. Of course, there is no Chinese wall separating theoretical from applied physics or theoretical mathematics from applied mathematics, to mention only two examples. Still, there is a difference that makes a difference. And this difference is precisely the degree of the explanatory power attained at the highest level of abstraction (theory) as compared to the lower level (interpretation).

Such a distinction is even more necessary, more indispensable, when we are talking about literary theory precisely because in general our training in the humanities does not give us that much knowledge concerning the analysis of concepts made in philosophy, nor does it give the specific training in scientific methodology. So we have to make do with what we have learned. This does not mean that we have to settle for what we are given.

The point is that there is a constant confusion (conflation even) of levels of what is strictly a scientific theory of narrative fiction *and* literary criticism or literary interpretation (idiosyncratic reading) of such and such a token of literary fiction or such and such a genre in the best-case scenario. Most so-called postclassical narratology, or so-called contextualized narratology, does not at all constitute a scientific theory of

---

4. For example, the analyses in *Colonialism and Cultural Identity* and *Understanding Indian Movies* are very culturally oriented.

narrative fiction. In the best-case scenario, postclassical narratology are limited applications of certain aspects of narrative theory.

Most work done at this level of applied literary criticism has not had any kind of impact on the development of a scientific theory of narrative fiction. The explanatory power and range of these applications and local interpretations is very small. The reason for this lack of explanatory power and impact is simply that most of the work published under this rubric is idiosyncratic interpretations of one narrative fiction or another; they are idiosyncratic in approach, in their method of analysis or interpretation, and in their findings. They cannot actually be taught to the next generation; the next generation can only become adept at mimicking the approaches. In the best-case scenario, what the students learn is to mimic a literary style, an idiosyncratic terminology, procedures that in actual fact do not lead to real verifiable generalizations. They are really local and idiosyncratic in their nature.

Of course, as I've said before, everybody can take their pleasure where they want. That said, perhaps this is a good place for us to distinguish clearly idiosyncratic from scientific approaches and outcomes. As we both know well, apples are not oranges.

PROBLEMS WITH HISTORICISM

PCH: In principle, historical sensitivity could oppose the sort of idiosyncrasy you are rightly criticizing. But for a moment, I would like to consider some of the problems with an approach to criticism that is entirely historicist and/or culturalist. To do that, it might be valuable to begin with the usual objection to cognitive literary study, an objection that I am sure you have heard as much as I have. When the speaker is polite, it has something like the following form: "Cognitive science may have some valuable insights. But don't you feel that it leaves out culture and history?" When asked questions of this sort, cognitivists—including myself—usually fumble about apologetically, saying that yes, some of us are guilty and we really need to do more to integrate culture into our cognitive analyses.[5]

But, on reflection, it seems that this gets things rather badly wrong. In fact, it is fairly rare to find cognitivist literary critics who simply ignore culture and history. The bias of the profession as a whole is such that

---

5. A version of the following arguments was presented at the Modern Language Association annual convention in 2011. I am grateful to Lisa Zunshine, the organizer of the session, and to the participants and audience members for their comments and questions [PCH].

one simply has to know certain things about, say, Shakespeare's time and place if one is going to be accepted as a critic of Shakespeare, if one is going to publish on Shakespeare, and so on. It simply is not possible, for example, to treat Shakespeare's language and references as if they were not historically specific. Of course, this is not to say that the cultural part of cognitive literary study is always done well. But the same point holds for the cultural part of cultural studies. For example, one could argue that an unfortunate amount of postcolonial criticism in the culture studies mode is somewhat tone-deaf to ethical, metaphysical, or other nuances in, say, South Asian cultural traditions. But the crucial point is that discursive and institutional constraints inhibit the degree to which criticism from any theoretical orientation can ignore history or culture.

In contrast, there seem to be no constraints whatsoever on the degree to which criticism can ignore neuroscience, cognitive and affective research, developmental studies, group dynamics, or any other forms of understanding that contribute to our sense of cross-cultural and transhistorical commonality. In other words, it simply is not the case that cognitivists grossly ignore culture and history. But it is commonly the case that culturalists grossly ignore cognitivist and related research.

FLA: My take on this problem is at once like-minded and different. I agree that the criticisms (culturalist or otherwise) leveled at those with a scientific bent are hugely imbalanced, irrational, limited and limiting. Largely, I've learned to do as the Dante-character's journey through the nine circles of hell: to look and move on. I do believe that the foundation of a scientific theory of narrative fiction must include a rigorous formulation of how story and discourse function. All sciences and, beyond that, all knowledge is *ancillary* to the scientific study of narrative fiction. From my point of view, all forms of knowledge are subordinate to the scientific theory of narrative fiction.

By this I mean that once I have developed an understanding of how narrative fiction works—and from this point of view, a key tool is narratology—I can enrich my interpretations of specific, concrete instances of narrative fiction artifacts. For instance, I consider that socio-neurobiology has a very important contribution to make to our understanding of the creative process involved in the production (authors) of the artifacts *and* their reception (readers). I consider that the same science has a huge contribution to make toward our understanding of how authors are able to create and how readers are able to *re-create* emotions and reactions. If we study how the affective or emotional system works, we can better understand how the author creates characters endowed with fictional (selected and selective) emotion systems and how they are re-created by the readers.

Undoubtedly, philosophy can teach us a lot about how the moral behavior of characters is invented by authors and perceived by readers. To this, we must add all sorts of knowledge. Quite often this includes our knowledge of history, current events (present politics, for instance), economics, and geography. We can say the same of cognitive science, developmental psychology, our knowledge of logic, mathematical thinking, the reasoning based on causality and counterfactual scenarios. All this can be very useful for the interpretation of particular works. However, none of these sciences separately nor all of them together constitute a scientific theory of narrative fiction.

This scientific theory that exists merely in embryonic form today can only develop if it reaches general or universal findings and conclusions. These findings have to be centered on the two pillars of narrative fiction, story and discourse. They have to be centered on how specifically these two pillars work, how the application of the generative operator of discourse yields an organic unity of subject matter and form in the creating of the fiction. And this not in regards to such and such a novel, film, story, film, or such and such an author of sixteenth-century Spain as opposed to France, say—but an understanding of how story and discourse work universally.

PCH: You certainly do not have to convince me of the value of studying universals. But even the recognition of universals requires cultural and historical knowledge. That is why I am stressing that the entire discussion of cognition and culture is topsy-turvy. Indeed, this leads us to ask what difference it would make if we tried to correct the problem by introducing cognitive study into culturalism. Here, we may briefly consider Gallagher and Greenblatt's well-known discussion of the potato debate, the debate over the use of potatoes as food for English workers. Gallagher and Greenblatt have discussed this topic in an illuminating way. Nothing I wish to say undermines their historical work. But some conclusions they draw from this seem problematic. Specifically, they show that a wide range of factors were involved in the debate and in the rejection of potatoes by English workers. They conclude from this that there is a "need for a thorough historicization of what counted as food and what felt like hunger" (125).

My point here is not to argue that this claim is entirely wrong. Gallagher and Greenblatt are certainly getting at something. But they have formulated their conclusion in a way that grossly overstates the case. As such, it seems more mistaken than correct. This is largely because they leave out any sense of what constitutes hunger.

FLA: Yes, it is true that they leave out important considerations in their approach to hunger—you mention physiological and cultural reasons. I would tread cautiously with Gallagher and Greenblatt for this and other reasons.

As you know well, the general principle of the philosophy of science should be kept centrally in mind. That is, no addition of examples of any phenomena will lead to the establishment of a science. No science is simply the sum of examples. No science is simply the sum of interpretations based on ideological or political considerations. No science is based on evidence that is made to fit the interpretation. And certainly no science can be based on evidence that is forged.

Nothing that I say here undermines *factually* Gallagher and Greenblatt's study. Yet, knowing that historians are not always the purest examples of intellectual probity, I also do not take their conclusions at face value. I believe that we have to go to the evidence and to many other historians and compare and contrast their accounts.

On the one hand there is what *is actually the case*—the matters of fact in the case of this famine, from the historical point of view. On the other hand, there are other matters such as the cultural and physiological. Once these facts are established, there remains the fact that the big genocidal famines have all been politically motivated, and executed for strictly political reasons.

PCH: It seems important to make some background points here—about emotion or motivation in general—before turning to hunger in particular. Emotion or motivation systems are largely subcortical systems, activated by a combination of endogenous and exogenous (body-internal and body-external) stimuli, which lead to subjective experiences of positive or negative hedonic tone and orient one toward behavior that will reduce aversive experiences or enhance pleasurable experiences. The basic operation of these systems is adaptive in that central cases of aversive experiences are harmful and central cases of pleasurable experiences are conducive to health or reproduction. For example, the fear system produces an aversive feeling and leads to flight from potentially dangerous situations. The different emotion systems may enhance or inhibit one another. For example, fear and attachment tend to be mutually inhibiting. Finally, there are connections between prefrontal cortex and emotion systems such that emotional response may affect inferential and related processing, but emotional response may also be modulated by such processing. For example, fear responses may guide one's long-term planning, but one's reflection on a situation may serve to inhibit one's fear response—perhaps one's aversive feeling or perhaps only one's behavior.

FLA: Nicely put, Patrick. Yes, there is undoubtedly the mind/brain system specialized in reasoning that may have an impact on the limbic system, just as the limbic system may have an impact on the reasoning system. This reciprocal mechanism is constantly at work through the intermediary of the blueprint that the author creates and that readers *reprocess* in their gap-filling activities.

As I have mentioned before, directors, authors, artists make a myriad of decisions in the course of producing a film, novel, or comic book, and in this, their emotive and reason systems are at work. The emotion and reason systems are part and parcel of these myriad decisions. They generate the blueprint out of decisions that always concern subject matter or shape. To a lesser degree, the same applies to the reader/audience's gap filling—an interpretation and constant ascription of meaning to whatever is being read, heard, or viewed. Thanks to this specific activity of the audience, the artifact is infused with life and acquires its existence as an aesthetic object.

PCH: Turning to hunger. As it happens, hunger is a surprisingly complex case. First, there appear to be two distinct but interrelated systems that are sometimes referred to as "hunger." (On these systems, see chapters 4 and 5 of Wong.) One is the taste-preference system. This is a system that bears on our hedonic relation to the experience of eating certain things. Like other motivation systems, this has both innate and acquired components. The innate component involves, for example, an evolved aversion to the taste of certain toxins. The acquired component is largely a matter of habituation to certain tastes, as well as the association of tastes with emotional memories, particularly in critical-period experiences. (These emotional memories might include, for example, the association of a particular food with food poisoning.)

The other relevant motivational system is that of feeding. Feeding behavior is a result of blood sugar regulation, gene expression related to amount of fatty tissue (see Gilliam, Kandel, and Jessell 47), and to some extent other factors such as the stimulation of stretch receptors in the stomach (see Wong 91). This system is closer to what we would ordinarily refer to as "hunger," as in the phrase "I feel hungry" or when we say "I need to eat something." In contrast, the taste-preference system is more relevant to such expressions as "I am hungry for a piece of cheesecake" or "I want something tasty." These two systems are probably mutually excitatory and mutually inhibitory (though the relation may be asymmetrical). In addition, both are partially inhibited by disgust, though disgust appears to have a stronger impact on taste preference than on feeding motivation.

FLA: As you remark, hunger has two components: the physiological and the social or cultural. Of course, there is a third, crucial component: the political component. The decision to inflict starvation on populations that results in genocide is a political move that has been made many times in history.

Famine is a very dreadful, terrible weapon used against the people. All famines are man-made. Famine is genocide in a very cruel way—killing thousands, sometimes millions, of people by inflicting a very slow and painful death on the people from children to elders.

This is the conclusion arrived at by Thomas Keneally in *Three Famines: Starvation and Politics*. He focuses on Ireland, Bengal, and Ethiopia. Today, the ongoing famine in Somalia is another example of this artificial political starving of large populations. There is tons of wheat and flour and food staples held in these huge UN warehouses. It is the warlords, by military means, who halt the distribution of these foods. There is today a totally artificially created famine, just as there was in Ethiopia. There are dozens of other examples of famines created for political purposes such as that created by Stalin in the Soviet Union. Mao Tse-tung also used this method in the 1950s. Starvation of the masses is an instrument used by rulers to control the population, to instill fear and pain massively among us.

There are other less drastic forms of manipulation of the production and distribution of food. After the 1910–17 Revolution in Mexico, the farmers owned the plots of land called "ahijados" that allowed for the self-sustenance of their families. The Mexican government then modified the constitution of the country and de facto made the *ahijados* disappear, turning them into private property and therefore commodities to be sold to the big U.S. corporations, a move that was accompanied by the establishing of a free-trade treaty with the United States. The end result: thousands of farmers became extremely impoverished and tried to solve the problem by immigrating to the United States or to the big cities in Mexico. This was what largely spurred on the rapid growth of both the legal and illegal migration of people from Mexico to the United States.

If we compare the figures from January 1, 1991, when NAFTA (the North American Free Trade Agreement) was signed and today, we can see that the migration has been massive. Why? Because people are literally starving in the countryside. Today, too, even when the border is more dangerous to cross than ever and the probability of being spotted and sent back (in the best of situations) is high, young and old, men and women still attempt to migrate in massive numbers because they are literally starving.

PCH: First, just to be clear, I am *not* saying that "hunger has two components: the physiological and the social or cultural." I am, rather, saying that there are two separate biological systems that might be referred to as "hunger." Like all motivational systems, these have innate components, critical-period components, and emotional-memory components. Some of the critical-period and emotional-memory components may (or may not) manifest differential patterns across cultures. I want to be careful to distinguish this view from the commonplace that there is always some biology and some culture—this idea is probably true, but it has no clear descriptive or explanatory value. Even worse than this commonplace is the common view that biology and culture are the only contributing factors—thus leaving aside individual experience, developmental patterns, group dynamics, physical constraints, and so on. Indeed, even differences in literary or artistic traditions are usually not a matter of "culture," as the term is ordinarily used. For example, the apparently unusual recurrence of parent–child separation and reunion plots in Japanese tradition probably results initially from the prominence of *The Tale of Genji* with its focus on such plots. It probably does not suggest some broader cultural pattern (e.g., parenting practices that foster attachment insecurity).

In any case, having noted the existence of different hunger systems, we may return to Gallagher and Greenblatt's claims, specifically their assertion that we need to historicize hunger. An understanding of human hunger motivation suggests the following. First, it is extremely unlikely that feeding motivation in general varies historically. There are some limited exceptions in extreme conditions. In cases of traumatic overarousal, neurological systems suffer damage. This is presumably true of the hunger system. Thus chronic hunger undoubtedly produces changes—almost certainly deleterious changes—in the operation of the feeding system. One may speak of this as "historical" if one wishes. But the real variable is not historical culture, but famine (whatever the ambient culture might be).

So, if English workers generally felt hunger in the same ways and for the same reasons as we do today (changes in blood glucose, decline in fatty tissue, etc.), what explains the rejection of the potato? A neurological model suggests that there must have been some sort of inhibition of the feeding system. This leads to three possibilities. First, there could be inhibition from taste preference. Second, there could be inhibition from the disgust system. Finally, there could be prefrontal inhibition.

Undoubtedly, there was some taste-preference inhibition. It seems that our emotion systems are calibrated slightly toward dispreference. For example, when faced with a stranger, our initial impulse is not trust but mild distrust or wariness (see Oatley, *Best* 73; for the more general point, see Zajonc 35). This is presumably the case for our taste and indeed feeding systems as well. Thus the lack of habituation to potatoes would produce a slight aversion. This is one implication of Gallagher and Greenblatt's argument that is unsurprising. It seems clear that taste preference is historically and culturally differentiated in certain respects.

Even more importantly, Gallagher and Greenblatt make a good case that the debates over the potato involved the invocation of repulsive imagery, particularly of dirt. Cognitively, this rhetoric would have the effect of associating potatoes with disgust-relevant emotional memories. Thus we have another form of inhibition that we would expect—the inhibition of both taste preference and feeding by an activated disgust system. Here, too, the conclusion is not terribly surprising.

FLA: Patrick, let me just say that I find Gallagher and Greenblatt's claim suspect simply from a common-sense perspective and knowing that historians (social, cultural, and so on) can be strongly guided by political aims. There have been many documented cases of historians deliberately destroying and deliberately misinterpreting documents. After the disintegration of the Soviet Union, many archives were open to general historians and other scholars. There was the British historian Robert Service, who had an all-expense-paid trip to Moscow and the KGB archives that contained a treasure trove of political documents—not to mention manuscripts of novels and poetry by Russian Soviet writers. In spite of this access, Service's biographies of Lenin and Trotsky have been proved to be complete fabrications. Many other so-called historians who had access to the KGB archives wrote articles and books with the sole purpose of furthering their political pet theories.

I am not sure we should take at face value Gallagher and Greenblatt's assertion that workers were willing to starve their families. We know that the Stalinist bureaucracy used famine—the deliberate slaughtering of cattle and burning of crops—to turn the people against the Bolsheviks.

Already, common sense and history point us to a different material reality than that based on Gallagher and Greenblatt's analysis—but of course, at the end of the day, I would need to do either the archival work itself or compare and contrast the facts excavated by historians along with their respective interpretations. Once the facts are established, I might further ask how the physiological and the cultural play a part in the use of famine as a political weapon—as a method of genocide. The long and the short of it is that scholarly titles and credentials do not make for competent seekers of the facts.

PCH: You raise valuable points. Among other things, it is important that we are dealing with abstract debates here, not a hungry person faced directly with a steaming plate of potato pancakes. But Gallagher and Greenblatt's analysis seems plausible for the conditions it addresses. Indeed, we really have not yet explained the rejection of the potato as outlined in their argument. For example, empirical studies show that the disgust system does not have strong inhibitory force on feeding behavior (see Hoefling and colleagues). At the very least, it would seem that the disgust system would have to be activated with something more than the mild aversion produced by unfamiliarity and some rhetorical imagery. This is where Gallagher and Greenblatt's analysis is most revealing. Or, rather, this is where their analysis could be revealing, if it were integrated with neuroscience of motivation (rather than being framed in terms of Saussurean "signifiers" [112]).

Specifically, Gallagher and Greenblatt suggest two other reasons for the rejection of the potato. First, there is anti-Irish feeling and the association of the Irish with the potato. It is well established that out-groups inspire anterior insula activation, which suggests disgust (see Fiske, Harris, and Cuddy). Gallagher and Greenblatt's argument indicates that group division can lead to intensified activation of the disgust system regarding foods that are distinctively associated with the out-group. If valid, the suggestion genuinely advances our understanding of human motivation systems as well as our understanding of intergroup relations. I know of no research on disgust that makes this connection. Thus it is an example of the ways in which historical and cultural study can advance cognitive knowledge. On the other hand, we can gain this insight only by recognizing that hunger should not be subjected to "thorough historicization." Rather, a partial historicization should be integrated with our understanding of human motivation systems.

The second point about Gallagher and Greenblatt's analysis is that not all the rejection is due to disgust. As their discussion makes clear, there were rational reasons to resist the food insecurity that tended to go along with reliance on the potato crop. This gives us a reason for the inhibition of feeding motivations that is not first of all subcortical but rather prefrontal. This too is remarkable. It suggests, for example, that at least some of the workers were led to inhibit potentially strong motivational impulses due to long-term planning. This is something that one rarely associates with masses. But it is implied by Gallagher and Greenblatt's research—though, again, this implication is recoverable only when that research is combined with neuroscientific work.

FLA: I think my position is clear on this score, so I'll simply state that we have to ask, will a cultural, historical, physiological approach to hunger tell us anything of importance about the political use of famine as a tool of genocide?

LITERATURE AND SOCIAL BETTERMENT

PCH: We might now turn to the issue of thematic criticism and the interpretation of political and ethical concerns in literary works. We might begin with a common view that literature improves us. This is an idea that achieved prominence in the Romantic period, but appears in different forms at different times and places. A recent view is that literature is "adaptive" in providing "offline" practice at resolving problems while not risking the harm that would result from actual, online activity. This idea is most famously associated with Steven Pinker (543), but it is quite widespread (see, for example, Tooby and Cosmides).[6]

We might stay with the adaptationist idea for a moment. It is undoubtedly the case that the cognitive process of simulation—hence concrete, causally constrained, hypothetical and counterfactual imagining—is adaptive. Moreover, literary narrative clearly involves simulation. The question is whether specifically literary simulation has any unique selective advantage or any selective advantage at all. The idea seems to me, well, ludicrous. Suppose I am thinking about going to a place for berries. I imagine going there, then imagine bears in the vicinity (recruiting relevant memories). I therefore avoid going for the berries and avoid being killed by bears. That clearly increases my chances of reproducing. (Recall that "adaptive" does not mean "good" or "smart" or "moral"; it just means providing a selective advantage in the passing on of the relevant genes.) But just how does reading *The Sound and the Fury* do this?

FLA: I think it is worth repeating that from my point of view so-called criticism or interpretation has no direct contribution to make to a scientific study of narrative fiction—and this is true not only with respect to the political or ethical but in general. The problem with so-called thematic criticism—whatever the orientation of the scholar—is that it is the kind of criticism that by its very nature *knows no boundaries*.

Let me explain in more detail. In my view, narrative fiction is not an imitative (mirroring) phenomenon. It is not a simulation either—at least not in your sense. It is not a dry run for real-life circumstances, emotions, and conflicts. Narrative fiction is not a *replica in any way of reality*.

Rather, I consider that narrative fiction is a creation. It is not something that reproduces reality but instead adds to reality. It is literally a creation. It always adds something new. This creation uses as building blocks any and all materials coming from reality. The author is free to use everything from the real world—another novel, a short story, a book of philosophy or history, an image of a child with soiled pants. From the point of view of thematics, nothing is off-limits. The author is a shameless pillager of everything out there in the world. The author uses these building blocks to provide the foundation for creating something new—an *addition to reality* and not a replica of reality, just as a chair, a table, a spear all *add* to reality.

6. For further discussion of evolution and literature, see chapter 8 of Hogan, *Cognitive Science* and "On the Origin."

Thematic analysis is limitless. Those that specialize in thematic criticism only select a handful of fields such as history, politics, economics, oppression, and so on. Thematic analysis is actually unmanageable. The few attempts at thematic analysis using a formal classification of themes as we see with Lévi-Strauss and Propp were so unconvincing that their work hasn't turned into a scientific research program.

I believe that for the relation to an object to be aesthetic it has to be a *nonutilitarian* relation. All the other relations we have with the world are in one way or another utilitarian; they seek specific goals. The aesthetic relation is the only relation we have with the world where we do not take the object as a *tool for anything*. I regard a bowl made by Cellini as a work of art. My relationship to this bowl is not a utilitarian relation: if I own one, I am not going to use it as a tool for eating my cereal in the morning. If I do and when I do, I cease to have an aesthetic relation with the object.

The aesthetic relation with the object is one of a nonutilitarian contemplation (a contemplation that here even goes beyond the Kantian sense of disinterested relation). The moment I establish a utilitarian relation with the object, I interrupt the aesthetic relation—and even eventually destroy it, if, for instance, I used the bowl as a hammer. The moment I go to *Absalom! Absalom!* to determine what the relations were between African Americans and Anglos during and after the Civil War, I have interrupted my aesthetic relation with *Absalom! Absalom!*

Thematic criticism does not acknowledge that narrative fiction is 1) a creation and not a reproduction of reality, and 2) an aesthetic object that acquires its status only within an aesthetic relation—a nonutilitarian relation. It is not an aesthetic object per se. Narrative fiction in its existence as part of material reality *creates an aesthetic situation in which the aesthetic relation can take place*. For instance, I write a novel as a work of art. The novel that I have just written, the material thing I have in my hands, *is not yet an aesthetic object* because there is *no aesthetic relation with it*. It is a material object that creates an aesthetic situation (your aesthetic event, I believe). It creates a situation that *may* lead to an aesthetic relation. This manuscript is a material reality that *opens up* the possibility of an aesthetic relation. I have created a new situation in the world in which there is a possibility of a new aesthetic relation of subject with this object newly created.

To sum up: The aesthetics is in neither the object nor the subject. It is not the relation between the object and subject. Since the object is not yet an aesthetic object, it represents a possibility for the realization of an aesthetic relation.

Thematic analysis is infinite because the building blocks of narrative fiction on which it rests are potentially infinite. Nobody can say at what moment creation will cease forever, so the potential number of aesthetic objects out there is limitless. For his aesthetic purpose in the writing of *Absalom! Absalom!* Faulkner chooses to focus on the interaction of African Americans, Anglos, and racially mixed characters, and therefore he builds his story from those blocks of reality. Since human relations are tinged with moral dilemmas, moral options, Faulkner ends up addressing morals in the novel as well as racial politics—the politics derived from the historical situation in the United States before and after the Civil War. But also among the building blocks, he makes omnipresent a wisteria vine, Southern architecture, and, of course, horses. The choice of morals and politics is a very obvious one. The fact itself of characters interacting in the story will involve (to different degrees of explicitness) moral relations among them as well as, very likely, political relations; inventing characters entails ascribing to them their social essence.

So a thematic approach that attends to race, class, gender, and so on in the novel is not a totally arbitrary choice, because of the prevalence of the moral and political dimension of narrative fiction. At the same time, there is a certain arbitrariness to this kind of focus since one could by the same token attend to botany in the novel, for instance.

I agree with you, Patrick, that narrative fiction does not edify. There are many examples of culturally refined people committing the worst atrocities. No music, poetry, or novel has stopped the butchers from slaughtering and the torturers from inflicting pain. The burden of the proof is on those who claim that narrative fiction edifies to give a plausible explanation as to why the so-called culturally refined and well-read can commit such atrocities. Nazi officers reciting passages from *Sorrows of Young Werther* (1774) all while committing acts of genocide.

The survival value, the evolutionary value, of narrative fiction is equally speculative. Let us assume that narrative fiction is found everywhere since the dawn of time. So, then, how much exposure to narrative fiction does one have to have in order for it to have an evolutionary value, and what kind of narrative fiction produces such value? Is it all kinds, or only certain kinds? We haven't identified a narrative fiction gene that is transmitted from generation to generation. Again, the burden of the proof is on the shoulders of those who make such claims, and not on ours. We do not have to bring empirical proof that they are wrong. They have to bring empirical proof to show us that they are right.

PCH: Just a point of clarification—simulation is a cognitive process that does not replicate the world but rather forms particular counterfactual or hypothetical trajectories of actions or events. The counterfactual or hypothetical simulation of these events is constrained by some principles of real-world activity. The constraints are to some extent based on memory (as we know from the process of induction and from recent neurological research [see Schacter, Addis, and Buckner]). There are presumably both innate and critical-period principles as well. Simulation involves, in part, perceptionlike imaginings that give rise to emotional/motivational responses.

In any case, often, the claims about benefits of literature concern its specifically moral influence. (The idea is ubiquitous. As such, it appears even among literary Darwinists; see for example Brian Boyd 197.) But there are at least two problems with this idea. First, it tends to treat the study of literature as a form of ethical training—often, in effect, propaganda.[7] Personally, I am not fond of the idea that we should be shaping the morals of our students.

Of course, it is also important not to misidentify what constitutes propaganda. This is the difficulty with the demand of the political Right that teachers be politically neutral. By "politically neutral," they in effect mean "political in a way that fits 'normal' ideas." It seems that a wide range of people (not solely conservatives) consider something to be political propaganda only when it does not conform to common opinions. In one standard view, it is not political propaganda to condemn terrorism simplistically as absolute evil, the mere incomprehensible desire to harm. In contrast, it is propaganda to discuss the reasons terrorists give for their actions and the conditions that motivate their actions. To take a literary example, it is not propaganda to discuss the display of Christian virtue in a medieval literary work. But it is propaganda to consider the ways in which one medieval work suggests lesbian sexuality or to critique another such work for misogyny.

In fact, it is a key political task of criticism to help students encode unnoticed political aspects of a work and to think about those aspects

---

7. Needless to say, not all views of the ethical benefits of literature necessitate a particular stand on specific ethical issues, still less the support of propaganda. For example, Schiller's general idea of aesthetic education and the training of sensibility does not seem to require particular ethical positions. The difficulty with such accounts (i.e., those that do not entail specific ethical stances) is different. It is often not clear that they isolate genuine ethical consequences at all.

critically. In this way, the actual situation is almost the precise opposite of the common view. It is precisely when criticism challenges common views that it is not propaganda. Nonetheless, this does not mean that we should set moral improvement as a primary task or as a task at all. Put differently, this only suggests that our criticism and interpretation should not contribute to the further moral and political degradation of our students.

FLA: I agree wholeheartedly, Patrick. I would, however, push us in the direction of foundations vis-à-vis pedagogical goals and the achieving of these goals. What am I to teach, and how do I teach? In the classroom I am, above all, a teacher. My code of conduct is *secular*. That is, I do not seek the moral or political education of our students *at all*. I seek to teach facts and how to interpret facts according to the best scientific methodology at hand. I teach students how they themselves can conduct research to discover facts. It is an education that seeks to eradicate as much as possible ideology and ideological considerations in the transmission of verifiable information and the research leading to the discovery of new verifiable information.

From this point of view, when we do literary criticism or interpretation in the classroom—not to be confused with the scientific study of narrative fiction—of such and such a novel or short story, for instance, or such and such a film or comic book, I can and I must discuss the most salient features of the text at hand. Among those features, very commonly there are political and moral ingredients that are worthwhile to analyze to see how such thematic features work within the aesthetic relation we have with the object.

Of course, other features are also salient. Often in our texts we need to excavate and then bring to the fore the emotion of love. This, along with many other emotions that are also very common (anger, hate, envy, grief, and so on), I along with my students submit to examination as they present themselves. Facts need interpretation. They do not need to be represented in terms of our own personal like and dislikes.

Students come to my classroom wanting to learn the method, approach, and findings of a scientific approach to narrative fiction—and not my opinions. My obligation when I sign up with the university is to be a teacher—that is, a secular teacher. So it is my obligation to give students what corresponds to a secular education.

PCH: Earlier, I mentioned two problems with the view that literature improves us. The second problem, already suggested, is that a lot of literature does just the opposite. It does not make us more human. Rather, it makes us more nationalistic, racist, sexist, ethnocentric, and so on. There is an entire branch of literary criticism devoted to isolating and evaluating these deleterious processes. It is called "ideological critique." Literary works operate to inculcate beliefs, to form in-group/out-group categories, to valorize internal group hierarchies (e.g., patriarchy), to support wars, and to do many other heinous things. Of course, not all literary works do this. Moreover, the sorts of work favored by literature professors tend to do this much less, and they even criticize common ideological tendencies. Nonetheless, the participation of narrative fiction in racist, sexist, classist, and other discourses seems undeniable.

Consider a book I recently got around to reading—Margaret Mitchell's enormously popular, Pulitzer Prize–winning novel, *Gone With the Wind*. It is in many ways a highly polished novel, worthy of high esteem as a literary accomplishment. Scarlett O'Hara is a magnificent character. Mitchell appropriates Romantic Satanism to create this smart, driven, amoral heroine who again and again proves herself the superior of all the men around her in intelligence, business acumen, ability to survive—and even in the ability to kill. Mitchell also takes up the common dichotomy of the good girl and the bad girl, presenting remarkable variations on that patriarchal theme. First, she repeats the division among women in Melanie and Scarlett—but, in doing so, she gives Melanie inner attitudes that are almost as strong as those of Scarlett. Second, and perhaps more significantly, she, so to speak, turns the tables on men by repeating the dichotomy with two men—Ashley Wilkes and Rhett Butler.

Yet the book is shocking for its racism. It celebrates the Ku Klux Klan as a group of high-minded men who bravely take on the national enemy to protect their women from rape. She repeatedly characterizes slavery as a benevolent institution, insisting that African resentment of slavery resulted only from the propaganda of the Yankees preying on the infantile intellectual capacities of Africans. She even appropriates Underground Railroad narratives to tell the outrageous tale of a former slave who escapes his employer in the North to return to the plantation where he was a slave. She presents the murder of Yankees, assertive Africans, and collaborationists—even outside the context of war—as permissible and at times highly laudable.

FLA: Most likely, by the time the book is chosen from the shelf, the reader is simply looking for confirmation of his or her opinion. I would need a mountain of evidence and support (causal and correlated) to convince me that a work of fiction (or works of fiction) can have the power to shape opinion, conviction—worldview. What a fantastic power that would be—infinitely more powerful than films or television and even more so than peers. (Teenagers seem to collect more ideological baggage from their peers than from parents or from the mass media they consume.) Whether for good or bad, it is investing works of literature with a power whose source is unexplained. Further, one would want to know how narrative fiction could end up having a much larger impact in establishing neurological connections in the brain than any other media.

In any event, the richness and complexity of the social, built, and natural environment is such that even if we had a steady diet of television or video games—as some kids do—the influence is still negligible.

PCH: Well, as I mentioned, there is empirical evidence for such effects. The case of *The Birth of a Nation*, cited by Prentice and Gerrig, seems directly comparable to Mitchell. In light of this research, we probably for the most part just want to avoid teaching a book such as *Gone With the Wind*. If we do teach it, we do not want to allow its propaganda to work on our students. Thus we want to consider the ways in which Mitchell radically distorts history, dehumanizes Africans, fosters a xenophobic Confederate nationalism, and so on. In doing this, it is important to consider not only facts and inferences, not only ideas, but also feelings. *Gone With the Wind* is a powerful book precisely because of the feelings it arouses. *Gone With the Wind* not only gets many of the facts wrong; perhaps even more importantly, it gets many of the feelings "wrong."

Here, we might consider a single point in the novel, critically examining how it works in contrast with a very different sort of narrative—Rabindranath Tagore's short story "Kabuliwallah." Tagore's story concerns a huge Afghani man named Rahamat who sells fruit in Calcutta. He forms an unusual friendship with a young girl, Mini, giving her fruit and laughing with her. Outside the tender relationship with this girl, he is rough and sometimes violent. At one point in the story, he is convicted of stabbing a man in a fight and sent to prison. On being released years later, he goes to Mini's home. It is her wedding day now, and Rahamat asks her father to give her a little present of fruit from him. The father tries to send Rahamat away, seeing his presence as inauspicious. But, in conversation, he learns that the man has a daughter of his own, the same age, whom he misses terribly. That is what underlay his relationship with Mini. Rahamat takes out a little piece of paper with the girl's handprint—he could hardly afford a photograph. For me, this is a deeply moving example of the way in which Tagore uses attachment relations to humanize characters who would otherwise be categorized as alien, as almost inhuman.

Now, we might consider the following scene from *Gone With the Wind*. A Yankee soldier arrives at Scarlett's home. It is during the Civil War (or war of secession or even war of independence, from the Confederate point of view). The Yankees have systematically brutalized the local people and are widely feared as terrorists. This man has a weapon, but he has put it away. Other than trespassing on her property, he has not done much. Simply being a Yankee is enough for Scarlett. She puts a gun to his face and blows his brains out (literally). She and Melanie then take his money and begin to go through his goods. One of the first things they find is "a miniature of a little girl in a gold frame" (436). For a moment, it looks as if they will be forced to recognize that Scarlett has killed a person, a man with a little girl back home. For a moment, it seems that Mitchell will do exactly what Tagore did, humanizing this out-group member by foregrounding his attachment relations. But this possibility is immediately undermined. The list of items in the knapsack becomes strange and inconsistent with this interpretation. Suddenly, Melanie realizes that the man stole all these things (437). He thereby becomes further dehumanized—and his murder seems more fully justified—because he not only lacks attachment relations himself; he has violated the attachment relations of others by stealing the child's picture simply out of greed for the gold frame.

FLA: If *Gone With the Wind* (1936) does not contaminate professors (readers) like us, I wonder if it likewise would not contaminate students—or anyone else for that matter.

The vast majority of the population in the most developed of countries really does not read. Literature as such really does not have much of an impact on the population. Not to mention that the majority of fiction out there is lazy fiction; the bricks of narrative taken from moral, political, emotional reality out there are those of the common doxa—the dominant ideology. In book after book after book we come across the same worldviews, emotion systems, and the like.

Of course, students have very little time for their studies, so we want to teach narrative fiction that is rich in its total aesthetic system. This would be my criterion for teaching *Absalom! Absalom!* and not teaching *Gone With the Wind*—a thick book that would require a lot of reading time and that would become boring rather quickly. It is the aesthetic value that I use to make my selection.

In actual practice, these days, I teach short stories high in aesthetic content and shape. This way, the students can learn a lot about narrative devices and the treatment of a whole variety of subject matter. Such stories demand a great concentration of narrative devices and plot to hit their target (their intended effect on the reader) with as few means and as quickly as possible. In a few pages or even in only a paragraph or a page, these short narrative fictions present a rich array of worldviews and moral options. They reveal much about how, for instance, U.S. Latino and Latin American authors create vivid fictional worlds (settings, geographies, chronologies, durations, and speeds) inhabited by interesting characters depicted with psychological complexity and immersed in moral dilemmas.

PCH: I do not wish to end by indicating that this means works such as *Gone With the Wind* are simply bad, politically or aesthetically. Given the levels of racism in the United States today, given the degree to which white people particularly are prone to dehumanize blacks, I do feel that it is probably better not to encourage anyone to read this novel (especially if it would mean *not* reading something else, such as Faulkner). However, there are many things that one can learn from Mitchell's book—ethical and political things—if one understands "learning" in an appropriately complex way. Indeed, a work such as *Gone With the Wind* suggests some of the problems with some common approaches to political criticism in literary study.

To put the matter as baldly as possible, I believe that most liberal Americans will learn much more about colonialism from reading *Gone With the Wind* than from reading many postcolonial works, including some of the most famous. Speaking personally, I found the novel revelatory; it suggested to me what many Southern whites must have felt about the Civil War, about the Union, and about Reconstruction. I was raised to believe unreflectively that the Civil War was a "good" war. It ended slavery, surely one of the great abominations of human history, and thus stood with World War II and its (tragically late) ending of the Holocaust. I had some sense that the Civil War was also a war of colonial conquest. But it was only in reading *Gone With the Wind* that I came to have a real sense of the human loss and material deprivation suffered by the majority population in the Southern states during and after the war. The novel chronicles the physical destruction of Atlanta, the brute insensitivity of invading troops (or at least what was seen as brute insensitivity), the burning of homes and crops, the loss of lives, the mutilation of bodies, the devastation of society.

The incident with the Yankee soldier is therefore important in two ways. First, as we have seen, it shows Mitchell's ability to dehumanize out-group members. Second, and no less important, it shows her ability to represent the rage born from suffering and humiliation. Given what they have gone through, how could Scarlett and Melanie think any differently about the Union soldier? One imagines that there are thousands of Scarletts and Melanies in Iraq, Afghanistan, and Pakistan today. Indeed, her account of the development of the Ku Klux Klan would seem relevant to a range of terrorist organizations.

Of course, the racial attitudes of Scarlett, Melanie, and Mitchell herself (given what one can infer from the novel) are heinous. But that is important too. Societies all have prejudices and internal violence. That does not make war against those societies any less colonialist, any less violent and destructive itself. Indeed, it suggests once again the importance of finding solutions to problems that avoid violence, as well as the associated humiliation. I wonder if part of the current political discrepancy between, say, New England and the South is the result of a historical legacy that, in the South, associates Northern liberalism with cruelty and dehumanization. Of course, I do not at all wish to compare the suffering of whites during and after the Civil War to the suffering of slaves or to that of postemancipation African Americans. Nor do I wish to suggest that Northern prejudices against Southerners, however demeaning, are in any way comparable to the attitudes of Northern and Southern whites toward blacks. However, it may be that we would all be better off if we recognized the colonialist elements in North–South relations and the partial truths in a work such as *Gone With the Wind*—while, of course, not losing sight of its more serious falsities as well.

FLA: I agree with your ethical and political criticism of the novel and the comparison you make between the novel and contemporary situations. However, if I were to study the novel in the classroom, I would provide enough contextual (historical, social, cultural) information for the students to connect the dots between past and present and focus the majority of the time sleuthing out how the generative devices work in the shaping of the story. That is, I would try to be consistently secular in my approach, analysis, and discussion with my students.

POLITICS AND MOTIVATION

PCH: This leads to a final topic. A fundamental implication of work in affective science is that there is no action without motivation and thus no action without the engagement of some motivational system. I have often had the thought that one difference between the Right and Left in this country is that the Right has its eye on actual material benefits while the Left wants to have a sense of moral superiority. This is, of course, overly simple. People across the political spectrum want both material advantage and self-esteem. Nonetheless, I wonder if there is some validity to my intuitive (over)generalization.

Consider affirmative action. The "conservative" position on the issue is that it involves prejudice against whites. Whites who oppose affirmative action tend to want, say, medical school admissions for themselves, for their relatives, and so on. In contrast, the arguments one hears from liberal whites in support of affirmative action tend to take the form of assertions about the need to give blacks an opportunity to catch up from their disadvantaged starting point. In a sense, liberal whites accept the view that affirmative action does lead to prejudice against whites. But they accept this prejudice in part because it gives them a feeling of moral purity.

One might ask at this point just what the other options are. One option is recognizing that in-group/out-group divisions strongly bias our evaluations. Even arbitrary assignment to groups affects people's evaluation of in- and out-group members' work (Duckitt 68–69). In consequence, we would expect hiring and admissions practices to be skewed toward in-group preferences. A minimal affirmative action program would be a matter of compensating for this biasing effect, which undermines democratic equality in civil society and, ultimately, in governance as well. For example, it might tie hiring to the demographics of the qualified applicant pool over a particular period. There are obviously difficulties with any such approach (e.g., due to issues of constitutionality). The point is that something along these lines is unlikely to be considered anyway because it does not appeal to the motivations of either the Right or the Left.

FLA: I would come at this from a slightly different position, Patrick. Oppression and exploitation in societies worldwide are becoming more and more evident. It is becoming more and more difficult to simply use ideology to try to hide this reality or to justify it. People everywhere are more and more determined to end this situation. It is an awareness that we see in the Middle East, Europe, Northern Africa, Wisconsin, Ohio, and California, as well as in Mexico and Chile. However, people's strength is not just in their numbers. They have to create their own organizations in their daily struggles. They have to be always extremely vigilant to ensure that their organizations remain representations of *their* interests and stop the ruling classes from taking over those organizations.

There is one form of organization of the people that is remarkable because it has, since it appeared in the late nineteenth century, constantly reappeared in all parts of the world where the struggle of the people has reached a high intensity and level of organization. It first emerged in 1871 under the name of the Paris Commune, reemerging over time under different names: workers councils, soldiers councils, and peasant councils in Russia (universally known as soviets).

What is characteristic about these organizations is that they are the most democratic form of organization there is. Those members of the councils that represent the people who are mobilized and struggling and fighting on behalf of their own interests are elected directly in the workplaces via a mandate or series of orders. They are literally *representing* the workers and fulfilling their mandates. In this capacity, they can be removed at a moment's notice by the workers they represent if, for whatever reason, the workers do not consider that the mandate has been carried out. There is immediate, direct accountability in the councils. All workers of the workplace are eligible to be representatives of the workers. Representation is never a privilege. You continue to earn a regular salary and return to the workplace when you are replaced by another representative. Nobody becomes a professional representative, and no one can be a representative without a mandate, obligation to the people who elected him or her. This is a real democratic form of organization and representation. We saw this recently developing in the central square of Tunis, Tunisia. Together with this form of organization the workers have developed their political organizations—their political parties.

Together with the struggle of the creation of councils, there has been the struggle to create trade unions representing exclusively the economic interests of the workers. Unfortunately, as we know, this has rarely been the policy of trade unions. As we know historically, unfortunately, the leadership of these parties has systematically passed to the enemy and has led the organizations to a policy of supporting the policies of the ruling classes.

A huge obstacle in the mobilization of the working people remains the leadership of trade unions and labor parties—organizations that do not possess at all the virtues that we find in workers councils; that is, the virtues of radical democracy.

In my mind I do not really see or analyze political problems in terms of Right and Left or in terms of which of the two major parties in the United States is more democratic. In my view of things, the basic and most fundamental forces acting in society are today centered in the working class (working people generally) and the building of their independent organizations based on true democracy: this is the form of organization we see with the workers councils.

This is our present and future. If we are to survive and not be destroyed by the huge arsenal that will wipe us from the face of this planet, and if we are to save the planet, undoubtedly the interest of the working people and the democratic forms of the working people must prevail and become the norm of all political representation the world over.

PCH: Very good points. I certainly do not wish to confine politics to governmental programs. However, in-group/out-group divisions and biases often require the establishment of compensatory policies even in the sorts of groups you mention.

I'll end with a suggestion that is perhaps less "suspicious," as Ricoeur might say (referring to a form of interpretation that sets out to dissipate illusions [see chapter 2 of his *Freud*])—or, more simply put, a suggestion that is more optimistic. We have already noted the evolutionary importance of simulation. As already noted, I have argued that there has to be some pleasure associated with simulation or we would never engage in aversive imaginings, such as imagining being attacked by bears when going to pick berries.[8] Simulation would lose its adaptive value if we simply avoided such aversive imaginings.

By the same token, however, simulation would lose its adaptive value if there were not some reward preference for truth. If we equally enjoyed true and false simulations, simulation would have no beneficial consequences. In short, we should have some sort of enhanced hedonic response to truth. This is at least consistent with our often strong aversive response to deception, even in cases where the deception has no practical consequences for our well-being. In interpersonal relations, this is bound up with another, related set of emotional responses—trust and distrust. Most significantly, it is involved with the "survival-related responses to

---

8. See *What Literature* 29.

anticipation" involving endogenous rewards for "correct predictions," as Vuust and Kringelbach put it (266)—and for learning, as I added earlier.

So, here is the suggestion. Some social critics are not motivated primarily by a striving for enhanced self-esteem (or wealth and power). Rather, they are motivated by a strong aversive response to falsity and an enhanced reward-system response to veracity. (Like other emotional responses, the degree of hedonic response to truth no doubt varies individually.) Of course, this does not mean that they always get things right. Anyone can make mistakes. Sometimes social critics of this sort make a lot of mistakes. But that does not mean the motivation is not one of seeking truth.

This seems to me the case with a writer such as Noam Chomsky. Personally, I do not believe that Chomsky's social criticism is motivated by self-esteem considerations. (Of course, being widely considered one of the greatest minds ever may diminish the degree to which one needs to bolster one's self-esteem.) Indeed, much of the power of his political criticism results from its different motivational source. Specifically, Chomsky seems to me to be someone with an unusually strong aversive response to falsity and an unusually strong reward response to truth.

FLA: Yes, some social critics are definitely inclined to favor truth and even to situate truth among the highest values such as fairness, equality, beauty, for instance. Of course, this does not just apply to certain social critics such as Noam Chomsky and before him Bertrand Russell—both strongly motivated by the attainment of truth.

I would say that truth is what motivates the work of *most* scientists and *some* scholars in the humanities as well. Of course, the concept of truth only applies to assertions and therefore linguistic expressions and mathematical formulae. To the world of art as a whole, the concept of truth only applies to the arts that involve the use of language, such as written narrative fiction or filmic or comic book narrative fiction, and so on.

We find truth, for instance, in certain assertions within fiction, when those assertions refer to building blocks of reality that have been incorporated into the building of the fiction. So truth in fiction when it is there and can be there is always a derivative truth. It is a truth based on the assertions concerning the materials taken from the real world.

Together with the search for truth there is a search for effective reasoning and effective methods of reasoning. So we have, along with the aspiration of discovering truth, the aspiration to create scientific explanations that exhibit internal coherence, balance, and a structure based on the most economical explanations: the principle of Occam's razor and the principle of parsimony, for instance. These notions are both scientific and aesthetic. That is, they are notions that point to beauty and that have certain affinities with beauty while being at the same time fundamental ingredients of scientific knowledge—of science in general.

WORKS CITED AND SUGGESTIONS FOR FURTHER READING

Abbott, H. Porter. *The Cambridge Introduction to Narrative*. Cambridge: Cambridge University Press, 2002.
Aldama, Frederick Luis. "Characters in Comic Books." In *Characters in Fictional Worlds: Understanding Imaginary Beings in Literature, Film, and Other Media*, ed. Jens Eder, Fotis Jannidis, and Ralf Schneider. Berlin: De Gruyter, 2010. 318–28.
———. "Ethnicity." In *Teaching Narrative Theory*, ed. David Herman, Brian McHale, and James Phelan. New York: Modern Language Association of America, 2010. 252–65.
———. *Formal Matters in Contemporary Latino Poetry*. Palgrave Macmillan, 2013.
———. "Narrative, Scientific Approaches." In Hogan, *Cambridge* 542–43.
———. *Spilling the Beans in Chicanolandia: Conversations with Writers and Artists*. Austin: University of Texas Press, 2006.
———, ed. *Toward a Cognitive Theory of Narrative Acts*. Austin: University of Texas Press, 2010.
———. *A User's Guide to Postcolonial and Latino Borderland Fiction*. Austin: University of Texas Press, 2009.
———. *Why the Humanities Matter: A Common Sense Approach*. Austin: University of Texas Press, 2008.
———. *Your Brain on Latino Comics*. Austin: University of Texas Press, 2009.
Anderson, Michael. "Incidental Forgetting." In Baddeley, Eysenck, and Anderson 191–216.
———. "Motivated Forgetting." In Baddeley, Eysenck, and Anderson 217–44.
Āraṇya, Harihārānanda. *Yoga Philosophy of Patañjali*. Trans. P. N. Mukerji. Albany, NY: State University of New York Press, 1983.
Arsalidou, Marie, Emmanuel Barbeau, Sarah Bayless, and Margot Taylor. "Brain Responses Differ to Faces of Mothers and Fathers." *Brain and Cognition* 74.1 (2010): 47–51.

Aziz-Zadeh, Lisa, and Antonio Damasio. "Embodied Semantics for Actions: Findings from Functional Brain Imaging." *Journal of Physiology—Paris* 102.1–3 (2008): 35–39.

Aziz-Zadeh, Lisa, Stephen Wilson, Giacomo Rizzolatti, and Marco Iacoboni. "Congruent Embodied Representations for Visually Presented Actions and Linguistic Phrases Describing Actions." *Current Biology*, September 19, 2006, 1818–23.

Baddeley, Alan. "What Is Memory?" In Baddeley, Eysenck, and Anderson 1–17.

Baddeley, Alan D., Michael W. Eysenck, and Michael C. Anderson. *Memory*. New York: Psychology Press, 2009.

Banfield, Ann. *Unspeakable Sentences: Narration and Representation in the Language of Fiction*. Boston; London; Melbourne & Henley: Routledge & Kegan Paul, 1982.

Bechara, Antoine, Hanna Damasio, Daniel Tranel, and Antonio Damasio. "Deciding Advantageously Before Knowing the Advantageous Strategy." *Science* 275.5304 (1997): 1293–94.

Beeman, Mark. "Coarse Semantic Coding and Discourse Comprehension." In Beeman and Chiarello 255–84.

Beeman, Mark, and Christine Chiarello, eds. *Right Hemisphere Language Comprehension: Perspectives from Cognitive Neuroscience*. Mahwah, NJ: Lawrence Erlbaum, 1998.

Bolaño, Roberto. *Between Parentheses: Essays, Articles, and Speeches, 1998–2003*. New York: New Directions, 2011.

Bower, G. H. "Affect and Cognition." *Philosophical Transactions of the Royal Society of London*, Series B 302 (1983): 387–402.

Boyd, Brian. *On the Origin of Stories: Evolution, Cognition, and Fiction*. Cambridge, MA: Harvard University Press, 2009.

Boyd, Julian C. "The Semantics of Modal Verbs." *Journal of Linguistics* 5 (1969): 57–74.

Brecht, Bertolt. *Die Stücke von Bertolt Brecht in Einem Band*. Frankfurt, Germany: Suhrkamp, 1992.

Brooks, Peter. *Reading for the Plot: Design and Intention in Narrative*. New York: Vintage, 1984.

Bunge, Mario. *Philosophy in Crisis: The Need for Reconstruction*. Amherst, NY: Prometheus Books, 2001.

Calvino, Italo. *Se una notte d'inverno un viaggiatore*. Cles, Italy: Mondadori, 1994.

Calvo-Merino, Beatriz, Corinne Jola, Daniel E. Glaser, and Patrick Haggard. "Towards a Sensorimotor Aesthetics of Performing Art." *Consciousness and Cognition* 17 (2008): 911–22.

Changeux, Jean-Pierre. *The Physiology of Truth: Neuroscience and Human Knowledge*. Cambridge, MA: Belknap Press of Harvard University Press, 2004.

Chatman, Seymour. *Coming to Terms: The Rhetoric of Narrative in Fiction and Film*. Ithaca, NY: Cornell University Press, 1990.

———. *Story and Discourse: Narrative Structure in Fiction and Film*. Ithaca, NY: Cornell University Press, 1978.

Chomsky, Noam. *Aspects of the Theory of Syntax*. Cambridge, MA: MIT Press, 1965.

———. *On Language: Chomsky's Classic Works* Language and Responsibility *and* Reflections on Language *in One Volume*. New York: The New Press, 1998.

———. *Syntactic Structures*. The Hague: Mouton, 1971.

Chiarello, Christine. "On Codes of Meaning and the Meaning of Codes: Semantic Access and Retrieval Within and Between Hemispheres." In Beeman and Chiarello 141–60.

Chiarello, Christine, and Mark Beeman. "Commentary: Getting the Right Meaning from Words and Sentences." In Beeman and Chiarello 245–51.

Churchland, Paul. *The Engine of Reason, The Seat of the Soul: A Philosophical Journey into the Brain*. Cambridge, MA: MIT Press, 1995.

Churchland, Paul, and Patricia Churchland. *On the Contrary: Critical Essays, 1987–1997*. Cambridge, MA: MIT Press, 1998.

Clore, Gerald L., and Andrew Ortony. "Cognition in Emotion: Always, Sometimes, or Never?" In *Cognitive Neuroscience of Emotion*, ed. Richard D. Land and Lynn Nadel with Geoffrey L. Ahern, John J. B. Allen, Alfred W. Kaszniak, Steven Z. Rapcsak, and Gary E. Schwartz. Oxford: Oxford University Press, 2000. 24–61.

Cohn, Dorrit. *The Distinction of Fiction*. Baltimore: Johns Hopkins University Press, 1999.

———. *Transparent Minds: Narrative Modes for Presenting Consciousness in Fiction*. Princeton, NJ: Princeton University Press, 1978.

Damasio, Antonio. *Descartes' Error: Emotion, Reason, and the Human Brain*. New York: Putnam, 1994.

———. *Looking for Spinoza: Joy, Sorrow, and the Feeling Brain*. Orlando, FL: Harcourt, 2003.

Dawson, Michael. *Connectionism: A Hands-On Approach*. Malden, MA: Blackwell, 2005.

Dehaene, Stanislas. "Fondements cognitive de l'arithmétique élémentaire." Lecture series, Collège de France, 2008. Available at www.college-de-france.fr.

Dehaene, Stanislas, Jean-Pierre Changeux, and Lionel Naccache. "The Global Neuronal Workspace Model of Conscious Access: From Neuronal Architectures to Clinical Applications." In *Characterizing Consciousness: From Cognition to the Clinic? Research and Perspectives in Neurosciences*, ed. Stanislas Dehaene and Yves Christian. Berlin: Springer-Verlag, 2011. 55–84.

Deutsch, Max. "Meaning Externalism and Internalism." In Hogan, *Cambridge* 472–74.

Doherty, Martin. *Theory of Mind: How Children Understand Others' Thoughts and Feelings*. New York: Psychology Press, 2009.

Dostoyevsky, Fyodor. *Crime and Punishment*. Trans. Sidney Monas. New York: New American Library, 1968.

Duckitt, John. *The Social Psychology of Prejudice*. New York: Praeger, 1992.

DuPlessis, Rachel Blau. "Manifests." *Diacritics* 26.3–4 (1996): 31–53.

Edelman, Gerald. *Second Nature: Brain Science and Human Knowledge*. New Haven, CT: Yale University Press, 2006.

Eysenck, Michael. "Semantic Memory and Stored Knowledge." In Baddeley, Eysenck, and Anderson 113–35.

Fabb, Nigel, and Morris Halle. *Meter in Poetry: A New Theory*. Cambridge: Cambridge University Press, 2008.

Fauconnier, Gilles, and Mark Turner. *The Way We Think: Conceptual Blending and the Mind's Hidden Complexities*. New York: Basic Books, 2002.

Faulkner, William. *The Sound and the Fury*. Ed. David Minter. 2nd ed. New York: Norton, 1994.

Fiske, Susan T., Lasana T. Harris, and Amy J. C. Cuddy. "Why Ordinary People Torture Enemy Prisoners." *Science*, November 26, 2004, 1482–83.

Fludernik, Monika. *Towards a "Natural" Narratology*. London: Routledge, 1996.

Fodor, Jerry. *Language of Thought*. Cambridge, MA: Harvard University Press, 1975.

———. *LOT 2: The Language of Thought Revisited*. New York: Oxford University Press, 2008.

Forgas, Joseph P. "Affect and Information Processing Strategies: An Interactive Relationship." In *Feeling and Thinking: The Role of Affect in Social Cognition*, ed. Joseph Forgas. Cambridge: Cambridge University Press, 2000. 253–80.

———, ed. *Feeling and Thinking: The Role of Affect in Social Cognition*. Cambridge: Cambridge University Press, and Paris: Editions de la Maison des Sciences de l'Homme, 2000.

Foucault, Michel. *The Archaeology of Knowledge and The Discourse on Language*. Trans. Alan M. Sheridan-Smith. New York: Harper and Row, 1972.

Franklin, Robert, and Reginald Adams. "The Two Sides of Beauty: Laterality and the Duality of Facial Attractiveness." *Brain and Cognition* 72.2 (2010): 300–305.

Frijda, Nico. *The Emotions*. Cambridge: Cambridge University Press, 1986.
Frye, Northrop. *Anatomy of Criticism: Four Essays*. Princeton, NJ: Princeton University Press, 1957.
Gallagher, Catherine, and Stephen Greenblatt. *Practicing New Historicism*. Chicago: University of Chicago Press, 2000.
Gazzaniga, Michael, ed. *Cognitive Neuroscience: A Reader*. Malden, MA: Blackwell, 2000.
———. *Human: The Science Behind What Makes Us Unique*. New York: Ecco, 2008.
Genette, Gérard. *Narrative Discourse: An Essay in Method*. Ithaca, NY: Cornell University Press, 1980.
———. *Narrative Discourse Revisited*. Ithaca, NY: Cornell University Press, 1988.
Gerrig, Richard, and Deborah Prentice. "Notes on Audience Response." In *Post-Theory: Reconstructing Film Studies*, ed. David Bordwell and Noël Carroll. Madison: University of Wisconsin Press, 1996. 388–403.
Gibbs, Raymond. "Embodiment." In Hogan, *Cambridge* 273–74.
Gilbert, Daniel T. "How Mental Systems Believe." *American Psychologist* 46.2 (1991): 107–19.
Gilliam, T. Conrad, Eric Kandel, and Thomas Jessell. "Genes and Behavior." In *Principles of Neural Science*, ed. Eric Kandel, James Schwartz, and Thomas Jessell. 4th ed. New York: McGraw-Hill, 2000. 36–62.
Gopnik, Alison. *The Philosophical Baby: What Children's Minds Tell Us about Truth, Love, and the Meaning of Life*. New York: Farrar, Straus and Giroux, 2009.
Gopnik, Alison, Andrew N. Meltzoff, and Patricia K. Kuhl. *The Scientist in the Crib: Minds, Brains, and How Children Learn*. New York: William Morrow, 1999.
Graff, Gerald. "The Pseudo-Politics of Interpretation." In *The Politics of Interpretation*, ed. W. J. T. Mitchell. Chicago: University of Chicago Press, 1983. 145–58.
Harley, Heidi. "Thematic Roles." In Hogan, *Cambridge* 861–62.
Harris, Zellig. *A Grammar of English on Mathematical Principles*. New York: Wiley, 1982.
———. *Methods in Structural Linguistics*. Chicago: University of Chicago Press, 1951.
Hayes, Bruce. *Metrical Stress Theory: Principles and Case Studies*. Chicago: University of Chicago Press, 1995.
Hegel, G. W. F. *Aesthetics: Lectures on Fine Art*. Vol. 1. Trans. T. M. Knox. Oxford: Clarendon Press, 1975.
Hempel, Carl. "The Function of General Laws in History." In *Aspects of Scientific Explanation and Other Essays in the Philosophy of Science*. New York: Free Press, 1965. 231–44.
Herman, David. *Basic Elements of Narrative*. Malden, MA: Wiley-Blackwell, 2009.
———, ed. *The Emergence of Mind: Representations of Consciousness in Narrative Discourse in English*. Lincoln: University of Nebraska Press, 2011.
———. *Narrative Theory and the Cognitive Sciences*. Stanford, CA: CSLI, 2003.
———. *Story Logic: Problems and Possibilities of Narrative*. Lincoln: University of Nebraska Press, 2002.
Hoefling, Atilla, et al. "When Hunger Finds No Fault with Moldy Corn: Food Deprivation Reduces Food-Related Disgust." *Emotion* 9.1 (2009): 50–58.
Hogan, Patrick Colm. *Affective Narratology: The Emotional Structure of Stories*. Lincoln: University of Nebraska Press, 2011.
———, ed. *The Cambridge Encyclopedia of the Language Sciences*. Cambridge: Cambridge University Press, 2011.
———. *Cognitive Science, Literature, and the Arts: A Guide for Humanists*. New York: Routledge, 2003.
———. *Colonialism and Cultural Identity: Crises of Tradition in the Anglophone Literatures of India, Africa, and the Caribbean*. Albany: State University of New York Press, 2000.

———. *Empire and Poetic Voice: Cognitive and Cultural Studies of Literary Tradition and Colonialism*. Albany, NY: State University of New York Press, 2004
———. "Generative Poetics." In Hogan, *Cambridge* 337–39.
———. *How Authors' Minds Make Stories*. Cambridge: Cambridge University Press, 2013.
———. *Joyce, Milton, and the Theory of Influence*. Gainesville, FL: University Press of Florida, 1995.
———. "Literature, God, and the Unbearable Solitude of Consciousness." *Journal of Consciousness Studies* 11.5–6 (2004): 116–42.
———. *The Mind and Its Stories: Narrative Universals and Human Emotion*. Cambridge: Cambridge University Press, 2003.
———. *Narrative Discourse: Authors and Narrators in Literature, Film, and Art*. Columbus: The Ohio State University Press, 2013.
———. *On Interpretation: Meaning and Inference in Law, Psychoanalysis, and Literature*. Athens: University of Georgia Press, 2008 (1996).
———. "On the Origin of Literary Narratives and Its Relation to Adaptation." In *Arts: A Science Matter*, ed. Maria Burguete and Lui Lam. Singapore: World Scientific, 2011. 267–92.
———. "Palmer's Anti-Cognitivist Challenge." *Style* 45.2 (2011): 244–48.
———. "Sensorimotor Projection, Violations of Continuity, and Emotion in the Experience of Film." *Projections: The Journal for Movies and Mind* 1.1 (2007): 41–58.
———. "Structure and Ambiguity in the Symbolic Order: Some Prolegomena to the Understanding and Criticism of Lacan." In *Criticism and Lacan: Essays and Dialogue on Language, Structure, and the Unconscious*, ed. Patrick Colm Hogan and Lalita Pandit. Athens: University of Georgia Press, 1990.
———. *Ulysses and the Poetics of Cognition*. New York: Routledge, 2014.
———. *Understanding Indian Movies: Culture, Cognition, and Cinematic Imagination*. Austin: University of Texas Press, 2008.
———. *Understanding Nationalism: On Narrative, Identity, and Cognitive Science*. Columbus, OH: Ohio State University Press, 2009.
———. *What Literature Teaches Us About Emotion*. Cambridge: Cambridge University Press, 2011.
Holland, John, Keith Holyoak, Richard Nisbett, and Paul Thagard. *Induction: Processes of Inference, Learning, and Discovery*. Cambridge, MA: MIT Press, 1986.
Iacoboni, Marco. *Mirroring People: The New Science of How We Connect with Others*. New York: Farrar, Straus and Giroux, 2008.
Ingalls, Daniel. Introduction to *The Dhvānyaloka of Ānandavardhana with the Locana of Abhinavagupta*, ed. Daniel Ingalls. Trans. Daniel Ingalls, Jeffrey Masson, and M. V. Patwardhan. Cambridge, MA: Harvard University Press, 1990. 1–39.
Ingram, John. *Neurolinguistics: An Introduction to Spoken Language Processing and Its Disorders*. Cambridge: Cambridge University Press, 2007.
Ito, Tiffany, Geoffrey Urland, Eve Willadsen-Jensen, and Joshua Correll. "The Social Neuroscience of Stereotyping and Prejudice: Using Event-Related Brain Potentials to Study Social Perception." In *Social Neuroscience: People Thinking About Thinking People*, ed. John Cacioppo, Penny Visser, and Cynthia Pickett. Cambridge, MA: MIT Press, 2006. 189–208.
Jaén, Isabel, and Julien Jacques Simon, eds. *Cognitive Literary Studies: Current Themes and New Directions*. Austin: University of Texas Press, 2012.
Jakobson, Roman. *Language in Literature*. Cambridge, MA: Belknap Press of Harvard University Press, 1987.

Jauss, Hans Robert. *Toward an Aesthetic of Reception*. Trans. Timothy Bahti. Minneapolis: University of Minnesota Press, 1982.
Kahneman, Daniel, and Dale Miller. "Norm Theory: Comparing Reality to Its Alternatives." *Psychological Review* 93.2 (1986): 136–53.
Kane, Julie. "Poetry as Right-Hemispheric Language." *Journal of Consciousness Studies* 11.5–6 (2004): 21–59.
Kant, Immanuel. *Critique of Judgement*. Trans. John H. Bernard. London: Hafner Press, 1951.
———. *Critique of Practical Reason*. Trans. Lewis White Beck. Indianapolis, IN: Bobbs-Merrill Educational, 1956.
———. *Groundwork of the Metaphysic of Morals*. Trans. Herbert J. Paton. New York: Harper and Row, 1964.
Kaufman, James, and Robert Sternberg, eds. *The Cambridge Handbook of Creativity*. Cambridge: Cambridge University Press, 2010.
Kawabata, Hideaki, and Semir Zeki. "Neural Correlates of Beauty." *Journal of Neurophysiology* 91.4 (2004): 1699–1705.
Keen, Suzanne. *Empathy and the Novel*. Oxford; New York: Oxford University Press, 2007.
Keneally, Thomas. *Three Famines: Starvation and Politics*. New York: Public Affairs, 2011.
Keysar, Boaz, and Dale Barr. "Self-Anchoring in Conversation: Why Language Users Do Not Do What They 'Should.'" In *Heuristics and Biases: The Psychology of Intuitive Judgment*, ed. Thomas Gilovich, Dale Griffin, and Daniel Kahneman. Cambridge: Cambridge University Press, 2002.
Kiparsky, Paul. "On Theory and Interpretation." In *The Linguistics of Writing: Arguments Between Language and Literature*, ed. Nigel Fabb, Derek Attridge, Alan Durant, and Colin MacCabe. New York: Methuen, 1987. 185–98.
———. "The Role of Linguistics in a Theory of Poetry." In *Essays in Modern Stylistics*, ed. Donald Freeman. London: Methuen, 1981. 9–23.
Kringelbach, Morten, and Kent Berridge, eds. *Pleasures of the Brain*. Oxford: Oxford University Press, 2010.
Labov, William. *Language in the Inner City: Studies in the Black English Vernacular*. Philadelphia, University of Pennsylvania Press, 1972.
———. *Principles of Linguistic Change, 3 Volume Set*. Oxford, UK: Wiley-Blackwell, 2010.
Labov, William, and Joshua Waletzky. "Narrative Analysis." In *Essays on the Verbal and Visual Arts*, ed. June Helm. Seattle: University of Washington Press, 1967. 12–44.
Lakoff, George. *Don't Think of an Elephant! Know Your Values and Frame the Debate—The Essential Guide for Progressives*. White River Junction, VT: Chelsea Green, 2004.
———. *Whose Freedom? The Battle Over America's Most Important Idea*. New York: Farrar, Straus and Giroux, 2006.
Lakoff, George, and Mark Turner. *More than Cool Reason: A Field Guide to Poetic Metaphor*. Chicago: University of Chicago Press, 1989.
Langlois, Judith H., and Lori A. Roggman. "Attractive Faces Are Only Average." *Psychological Science* 1.2 (1990): 115–21.
Lasnik, Howard. "Minimalism." In Hogan, *Cambridge* 502–5.
LeDoux, Joseph. *The Emotional Brain: The Mysterious Underpinnings of Emotional Life*. New York: Touchstone, 1996.
———. *Synaptic Self: How Our Brains Become Who We Are*. New York: Viking, 2002.
Lévi-Strauss, Claude. *Myth and Meaning*. New York: Schocken Books, 1979.
Lukács, Georg. *History and Class Consciousness: Studies in Marxist Dialectics*. Trans. Rodney Livingstone. Cambridge, MA: MIT Press, 1983.

———. "Marx and the Problem of Ideological Decay." In *Essays on Realism*, ed. Rodney Livingstone, trans. David Fernbach. Cambridge, MA: MIT Press, 1981. 114–66.

Martindale, Colin, and Kathleen Moore. "Priming, Prototypicality, and Preference." *Journal of Experimental Psychology: Human Perception and Performance* 14.4 (1988): 661–70.

Marx, Karl, and Friedrich Engels. *The Communist Manifesto*. London; New York: Penguin Books, 2002.

Massey, Irving. *The Neural Imagination: Aesthetic and Neuroscientific Approaches to the Arts*. Austin: University of Texas Press, 2009.

Maturana, Humberto, and Francisco Varela. *Autopoiesis and Cognition: The Realization of the Living*. Boston: Reidel, 1980.

McHale, Brian. "Beginning to Think About Narrative in Poetry." *Narrative* 17.1 (2009): 11–27.

McLeod, Peter, Kim Plunkett, and Edmund Rolls. *Introduction to Connectionist Modeling of Cognitive Processes*. Oxford: Oxford University Press, 1998.

Mitchell, Margaret. *Gone With the Wind*. New York: Avon, 1973.

Mounier, Emmanuel. *Be Not Afraid: Studies in Personalist Sociology*. New York: Harper, 1956.

Nadal, Marcos, Enric Munar, Miquel Capó, Jaume Rosselló, and Camilo Cela-Conde. "Towards a Framework for the Study of the Neural Correlates of Aesthetic Preference." *Spatial Vision* 21.3–5 (2008): 379–96.

Nalbantian, Suzanne. "Neuroaesthetics: Neuroscientific Theory and Illustration from the Arts." *Interdisciplinary Science Reviews* 33.4 (2008): 357–68.

Nisbett, Richard, and Lee Ross. *Human Inference: Strategies and Shortcomings of Social Judgment*. Englewood Cliffs, NJ: Prentice-Hall, 1980.

Oatley, Keith. *Best Laid Schemes: The Psychology of Emotions*. Cambridge: Cambridge University Press, 1992.

———. *Emotions: A Brief History*. Malden, MA: Blackwell, 2004.

———. *Such Stuff as Dreams: The Psychology of Fiction*. Malden, MA: Wiley-Blackwell, 2011.

———. "Why Fiction May Be Twice as True as Fact: Fiction as Cognitive and Emotional Simulation." *Review of General Psychology* 3.2 (June 1999): 101–17.

Oatley, Keith, Dacher Keltner, and Jennifer Jenkins. *Understanding Emotions*. 2nd ed. Malden, MA: Blackwell, 2007.

Ortony, Andrew. "The Role of Similarity in Similes and Metaphors." In *Metaphor and Thought*, ed. Andrew Ortony. 2nd ed. Cambridge: Cambridge University Press, 1993. 342–56.

Palmer, Alan. "Social Minds in Fiction and Criticism." *Style* 45.2 (2011): 196–240.

Palmer, Linda. "Kant and the Brain: A New Empirical Hypothesis." *Review of General Psychology* 12.2 (2008): 105–17.

Panksepp, Jaak. "The Affective Brain and Core Consciousness: How Does Neural Activity Generate Emotional Feelings?" In *Handbook of Emotions*, ed. Michael Lewis, Jeannette Haviland-Jones, and Lisa Feldman Barrett. 3rd ed. New York: Guilford Press, 2008. 47–67.

Patañjali. In Āraṇya, 1-407.

Phelan, James. *Experiencing Fiction: Judgments, Progressions, and the Rhetorical Theory of Narrative*. Columbus: The Ohio State University Press, 2007.

———. *Living to Tell About It: A Rhetoric and Ethics of Character Narration*. Ithaca, NY: Cornell University Press, 2005.

Phelan, James, and Peter J. Rabinowitz. *A Companion to Narrative Theory*. Oxford: Blackwell, 2005.

Piera, Carlos. "Intonational Factors in Metrics." *Belgian Journal of Linguistics* 15.1 (2001): 205–28.
Pinker, Steven. *How the Mind Works.* New York: Norton, 1997.
Premcand, Muṁśī. *Godān.* New Delhi, India: Rupa, 2004.
Prentice, Deborah, and Richard Gerrig. "Exploring the Boundary between Fiction and Reality." In *Dual-Process Theories in Social Psychology,* ed. Shelly Chaiken and Yaacov Trope. New York: Guilford Press, 1999. 529–46.
Prince, Gerald. *A Grammar of Stories.* The Hague: Mouton, 1973.
——. *Narratology: The Form and Functioning of Narrative.* New York: Mouton, 1982.
Proust, Marcel. *À la recherche du temps perdu.* Vol. 1, *Du côté de chez Swann.* Paris: Gallimard, 1954.
Ramachandran, V. S. *The Tell-Tale Brain: A Neuroscientist's Quest for What Makes Us Human.* New York: Norton, 2011.
Reddy, Michael. "The Conduit Metaphor: A Case of Frame Conflict in Our Language About Language." In *Metaphor and Thought,* ed. Andrew Ortony. Cambridge: Cambridge University Press, 1993. 164–201.
Reisch, George. "Chaos, History, and Narrative." *History and Theory* 30.1 (1991): 1–20.
Ricoeur, Paul. *Freud and Philosophy: An Essay on Interpretation.* Trans. Denis Savage. New Haven, CT: Yale University Press, 1977.
Rosch, Eleanor. "Prototypes." In Hogan, *Cambridge* 680–82.
Russell, Richard. "Sex, Beauty and the Relative Luminance of Facial Features." *Perception* 32.9 (2003): 1093–1107.
Russom, Geoffrey. "Word Patterns and Phrase Patterns in Universalist Metrics." In *Frontiers in Comparative Prosody,* ed. Mihhail Lotman and Maria-Kristiina Lotman. Bern, Switzerland: Peter Lang, 2011. 337–71.
Said, Edward. *Orientalism.* New York: Pantheon, 1978.
Sappho. *Sappho: A New Translation.* Ed. and trans. Mary Barnard. Berkeley: University of California Press, 1958.
Sartre, Jean-Paul. *Being and Nothingness.* Trans. Hazel Barnes. New York: Washington Square, 1966.
Saussure, Ferdinand de. *Course in General Linguistics.* Trans. Wade Baskin. New York: Columbia University Press, 2011.
Schacter, Daniel. *Searching for Memory: The Brain, the Mind, and the Past.* New York: Basic Books, 1996.
Schacter, Daniel, Donna Rose Addis, and Randy L. Buckner. "Remembering the Past to Imagine the Future: The Prospective Brain." *Nature Reviews: Neuroscience* 8.9 (2007): 657–61.
Scheff, Thomas J. "Multipersonal Dialogue in Consciousness: An Incident in Virginia Woolf's 'To the Lighthouse.'" *Journal of Consciousness Studies* 7.6 (2000): 3–19.
Schiller, Friedrich. *On the Aesthetic Education of Man in a Series of Letters.* Trans. Reginald Snell. London: Routledge and Kegan Paul, 1954.
Shakespeare, William. *Hamlet.* Ed. Susanne Wofford. Boston: Bedford Books, 1994.
Sharvit, Yael. "The Puzzle of Free Indirect Discourse." *Linguistics and Philosophy* 31.3 (2008): 353–95.
Shaver, Phillip, and C. Hazan. "A Biased Overview of the Study of Love." *Journal of Social and Personal Relationships* 5.4 (1988): 473–501.
Shimamura, Arthur P., and Stephen E. Palmer, eds. *Aesthetic Science: Connecting Minds, Brains, and Experience.* Oxford: Oxford University Press, 2012.

Shklovsky, Victor. *Theory of Prose.* Trans. Benjamin Sher. Elmwood Park, IL: Dalkey Archive Press, 1991.
Skov, Martin. "The Pleasures of Art." In Kringelbach and Berridge 270–83.
Sperber, Dan, and Deirdre Wilson. *Relevance: Communication and Cognition.* Oxford, UK; Cambridge, MA: Blackwell, 1995.
Stapp, Henry. "Quantum Approaches to Consciousness." In *The Cambridge Handbook of Consciousness,* ed. Philip David Zelazo, Morris Moscovitch, and Evan Thompson. Cambridge: Cambridge University Press, 2007. 881–908.
Steen, Gerard. "The Paradox of Metaphor: Why We Need a Three-Dimensional Model of Metaphor." *Metaphor and Symbol* 23 (2008): 213-241.
Stein, Dan, and Bavanisha Vythilingum. "Love and Attachment: The Psychobiology of Social Bonding." *CNS Spectrums* 14.5 (2009): 239–42.
Sternberg, Meir. *Expositional Modes and Temporal Ordering in Fiction.* Baltimore: Johns Hopkins University Press, 1978.
Surányi, Balázs. "Merge." In Hogan, *Cambridge* 482–83.
———. "Principles and Parameters Theory." In Hogan, *Cambridge* 666–70.
Swirski, Peter. *Literature, Analytically Speaking: Explorations in the Theory of Interpretation, Analytic Aesthetics, and Evolution.* Austin: University of Texas Press, 2010.
Tagore, Rabindranath. "Exercise-Book." In *Selected Short Stories,* ed. William Radice. New York: Penguin, 2005. 140–45.
———. "Kabuliwallah." In *Selected Short Stories,* ed. William Radice. New York: Penguin, 2005. 113–20.
Tan, Ed. *Emotion and the Structure of Narrative Film: Film as an Emotion Machine.* Trans. Barbara Fasting. Mahwah, NJ: Lawrence Erlbaum, 1996.
Tooby, John, and Leda Cosmides. "Does Beauty Build Adapted Minds? Toward an Evolutionary Theory of Aesthetics, Fiction, and the Arts." *SubStance: A Review of Theory and Literary Criticism* 94/95 (2001): 6–27.
Trubetzkoy, Nikolai Sergeyevich. *Writings on Literature.* Minneapolis: University of Minnesota Press, 1990.
Turner, Frederick. "The Neural Lyre: Poetic Meter, the Brain, and Time." In *Natural Classicism: Essays on Literature and Science.* New York: Paragon House, 1985. 61–108.
Tversky, Amos. "Features of Similarity." *Psychological Review* 84.4 (1977): 327–52.
Vartanian, Oshin, and Vinod Goel. "Neuroanatomical Correlates of Aesthetic Preference for Paintings." *Neuroreport* 15.5 (2004): 893–97.
Veeser, H. Aram. Introduction to *The New Historicism,* ed. H. Aram Veeser. New York: Routledge, 1989, x–xvi.
Vermeule, Blakey. *Why Do We Care About Literary Characters?* Baltimore: Johns Hopkins University Press, 2009.
Villablanca, Jaime. "Why Do We Have a Caudate Nucleus?" *Acta Neurobiologiae Experimentalis* 70 (2010): 95–105.
Vuust, Peter, and Morten Kringelbach. "The Pleasure of Music." In Kringelbach and Berridge 255–69.
Vygotsky, L. S. *The Essential Vygotsky.* Ed. Robert W. Rieber and David K. Robinson. New York: Kluwer Academic/Plenum, 2004.
———. *Mind in Society: The Development of Higher Psychological Processes.* Ed. Michael Cole, Vera John-Steiner, and Sylvia Scribner. Cambridge, MA: Harvard University Press, 1978.
———. *Thought and Language.* Cambridge, MA: MIT Press, 1962.
Whitfield, T. W. Allan, and Philip Slatter. "The Effects of Categorization and Prototypicality

on Aesthetic Choice in a Furniture Selection Task." *British Journal of Psychology* 70.1 (1979): 67–75.

Woloch, Alex. *The One vs. the Many: Minor Characters and the Space of the Protagonist in the Novel.* Princeton, NJ: Princeton University Press, 2003.

Wong, Roderick. *Motivation: A Biobehavioural Approach.* Cambridge: Cambridge University Press, 2000.

Wurtz, Robert, and Eric Kandel. "Perception of Motion, Depth, and Form." In *Principles of Neural Science,* ed. Eric Kandel, James Schwartz, and Thomas Jessell. 4th ed. New York: McGraw-Hill, 2000. 548–71.

Zajonc, Robert B. "Feeling and Thinking: Closing the Debate Over the Independence of Affect." In Forgas 31–58.

Zeki, Semir. "Art and the Brain." *Daedalus* 127.2 (1998): 71–103.

———. *Inner Vision: An Exploration of Art and the Brain.* Oxford: Oxford University Press, 1999.

———. *Splendors and Miseries of the Brain: Love, Creativity, and the Quest for Human Happiness.* Chichester, UK; Malden, MA: Wiley-Blackwell, 2009.

Zunshine, Lisa, ed. *Introduction to Cognitive Cultural Studies.* Baltimore: Johns Hopkins University Press, 2010.

———. *Strange Concepts and the Stories They Make Possible: Cognition, Culture, Narrative.* Baltimore: Johns Hopkins University Press, 2008.

———. *Why We Read Fiction: Theory of Mind and the Novel.* Columbus: The Ohio State University Press, 2006.

INDEX

Abbott, H. Porter, 83
abduction, 7
aesthetics: beauty, 119–31, 135, 136, 140–48; goal, 50, 79, 83, 90, 95, 96, 97, 98, 99, 100, 108, 109, 112, 142, 145; history and art, 132–40; pleasure, 118–26; scientific approach, 116–48; sources of, 126–32
Aldama, Frederick: *Why the Humanities Matter,* 153
Allende, Isabel, 136–37, 138, 139, 142
American Revolution, 5
Amis, Martin: *Time's Arrow,* 100
analytical tool 26, 32, 37, 41, 50, 53, 54
Animal House, 141
Aristotle, 57, 146, 147, 154
Arquette, David: *Roadracers,* 94
art: and history, 132–40
Austen, Jane, 24
Aziz-Zadeh, Lisa, 76, 77

Ball, Alan: *True Blood,* 141
Balzac, Honoré de, 108–9, 111

Banfield, Ann, 35, 39
Barthes, Roland, 39; *S/Z,* 41
Baumgarten, Alexander Gottlieb: *Aesthetics,* 133
The Birth of a Nation, 157, 178
Bloomfield, Leonard, 35, 37
blueprint, 28, 34, 60, 79, 88, 89, 112, 131, 142, 143, 156, 166
Boaz, Franz, 35, 37
bodily cognition, 72, 73, 77
Bosch, Hieronymus: *Garden of Earthly Delight,* 141
Boyd, Julian, 35
brains, 7, 44, 54, 76, 95, 116, 120, 121, 124, 135, 144, 146, 178, 179; children, 21, 30; cognition, 72, 73, 74; computational spaces, 3; connectionism, 66–72; emotional system, 16; evolved, 11; executive, 16, 17, 45–46; filter, 17; mechanisms, 12, 13; mind emerging, 4; mirror neuron system, 15; neurochemical makeup, 2
Brecht, Bertolt, 147; *Die Maßnahme (The Measures Taken),* 111

Brooks, Peter, 42
Buddhism, 5
Bunge, Mario, 5; *Philosophy in Crisis*, 3
Bush, George W., 26, 27

Calvino, Italo, 188: *Se una notte d'inverno un viaggiatore*, 101
Capote, Truman: *In Cold Blood*, 83–84
Carpentier, Alejo: "Viaje a la semilla," 100
Carrington, Leonora, 141
Change, 39
Changeux, Jean-Pierre, 2, 4
character: round, 23, 24, 26
Chatman, Seymour, 54, 83, 149
childhood association, 23n10
child rearing, 7, 30
children, 64–65
Chomsky, Noam, 26, 37–38, 44, 64, 68–69, 154; grammatical theory, 63; and internalism, 52–59, 61, 65; *Language and Responsibility*, 39–40; "Minimalist Program," 55; social criticism, 185; *Syntactic Structures*, 49–50
Churchland, Patricia, 74
Churchland, Paul, 74
Clore, Gerald, 73
cognitive linguistics, 52, 54, 106; and embodied cognition, 72–78
coherence, 12, 13, 59, 153, 159, 186
coevolution, 8, 30
Cohn, Dorritt, 83, 91, 93
connectionism, 66–72
consciousness, 3, 4, 9, 10, 12, 12n5; levels, 56; stream, 48, 59
consequentialism, 5, 6, 12, 13, 14, 28, 29, 34, 39, 56, 63, 71, 75, 76, 77, 82, 92, 105, 110, 120, 127, 131, 160, 182, 184; adaptive, 124; ethics, 175n7
contradiction, 11, 35, 110, 111; task, 12, 12n5
Cortázar, Julio, 114; "The Continuity of Parks," 99
cortex, 12; anterior cingulate, 12n5, 70; executive, 17; lateral prefrontal, 12n5; motor, 3, 17, 76; neo, 116; parietal, 119; prefrontal, 17, 166; sensory, 135; somatosensory, 118

critical-period experience, 18, 20, 166, 168, 175
criticism: literary, 39–40, 86, 135, 156–62, 163, 177; political, 180, 181, 185; social, 185; thematic, 171, 172, 173, 176
culture, 6, 8, 27, 103–4, 154, 168; cognition, 164; domain of, 112; and history, 162–63, 169; intellectual, 104

Dalí, Salvador, 141
Damasio, Antonio, 72–73, 74, 76, 86
Darwinism, 8, 175
deduction, 7
Dehaene, Stanislas, 2–3
Deliverance, 99–100
deontology, 28, 29
Derrida, Jacques, 35, 39, 40, 153
Descartes, René, 12
descriptive linguistics, 53
determinism, 5, 112
dialogic speech, 62, 63, 64, 65
discourse analysis, 38–39, 40, 41, 56, 82
Dostoyevsky, Fyodor, 90, 96, 155; *The Brothers Karamazov*, 97; *Crime and Punishment*, 13, 97; *Demons*, 97
documentary tendency, 82, 83
Duchamp, Marcel, 136
Ducrot, Oswald, 40
DuPlessis, Rachel Blau: "Manifests," 102

Edelman, Gerald, 124
Ellison, Ralph: *Invisible Man*, 152
embedded theory of mind, 24–25, 50, 64
embodied cognition: and cognitive linguistics, 72–78
emotion, 80; and ethics, 113–15; reflex, 16, 17, 19; and self, 16–22. *See also* story emotion
empathy, 16, 21, 28, 30, 140
emplotment, 23, 27, 28, 84, 89, 92 106, 107, 132
Engels, Friedrich, 111; *Communist Manifesto*, 155
Enlightenment, 104, 146
Esprit, 9

ethics, 15, 26, 29, 33, 86, 104, 126, 127, 128, 155, 160, 163, 171, 172, 175, 175n7, 180, 181; and emotion, 113–15; virtue, 28
evolution, 7, 30, 57, 124, 172n6
evolutionary psychology, 7, 8
expertise, 23
externalism, 57, 61, 65, 66–67
Eysenck, Michael, 76

Fabb, Nigel, 38
fait divers, 84
fantasy, 22, 23
Fauconnier, Gilles, 106
Faulkner, William, 28, 48, 71, 90, 112, 145, 155, 180; *Absalom, Absalom!,* 26, 41, 174; *The Sound and the Fury,* 47, 50, 60, 61, 62, 64, 69, 77,159
Faye, Jean-Pierre, 39
fear, 8, 13, 17, 74, 154, 165–66
features of language, 49, 50
Fiennes, Ralph: *Coriolanus,* 25
firsthand experience, 33, 34
Flaubert, Gustave, 93, 97, 98, 155; *Madame Bovary,* 83, 84, 95, 114
Fodor, Jerry, 44, 56
Foucault, Michel, 82, 86, 87, 98, 149
Franklin, Benjamin, 5, 7
Franklin, Robert, 123
Friedrich, Caspar David: *Wanderer above the Sea of Fog,* 139
Fuentes, Carlos, 34, 112; *The Death of Artemio Cruz,* 34; *La region mas Transparente* (*Where the Air Is Clear*), 61

Gallagher, Catherine, 164, 165, 168–71
García Márquez, Gabriel, 84, 90, 137, 138, 139, 142; *One Hundred Years of Solitude,* 136
Gazzaniga, Michael, 11
Generation Kill, 156
generative linguistics, 39, 44, 52, 54, 63, 65, 69, 70, 83, 84, 90, 93, 96, 100, 103, 107, 113, 114, 115, 143, 150, 158, 164, 181
genetic determination, 5, 6

Genette, Gérard, 37, 54, 83, 91
genome, 7, 31
Géricault, Théodore: *Raft of the Medusa,* 140–41
Gerrig, Richard, 81, 156, 178
Gibbs, Raymond, 76–77
Gilbert, D. T., 156
goals, 13, 15, 19, 141, 151, 173; aesthetic, 50, 79, 83, 90, 95, 96, 97, 98, 99, 100, 108, 109, 112, 142, 145; long-term, 110; metaphor, 75; pedagogical, 176; science of language, 40, 54, 56; thematic, 59
Goethe, Johann Wolfgang von: *Faust,* 26; *West-Eastern Divan,* 26
Gone With the Wind, 177, 178, 179, 180, 181
Gopnik, Alison, 65; *Philosophical Baby,* 33; *Scientist in the Crib,* 33
Graff, Gerald, 86
"grasp," 73, 76–77, 116
Greenblatt, Stephen, 164, 165, 168–71

habituation, 18, 120, 124, 125, 136, 137, 138, 139, 141,143, 166, 169
Halle, Morris, 38
Harley, Heidi, 46
Harris, Zellig, 37
Hayes, Bruce, 38
HBO, 141
Hegel, G. W. F., 66, 132, 133
Hemingway, Ernest, 33–34, 90, 96, 155
Hempel, Carl, 114
Herman, David, 54; *Story Logic,* 46
hierarchical structures, 50, 55, 142, 177
historicism, 162–71
Hitler, Adolf: *Mein Kampf,* 26, 27, 28
Hjelmslev, Louis, 37
Hogan, Patrick Colm: *Cognitive Science, Literature, and the Arts,* 157; *Colonialism and Cultural Identity,* 161; *Joyce, Milton, and the Theory of Influence,* 138; *The Politics of Interpretation,* 35; *Understanding Nationalism,* 105n4; *What Literature Teaches Us About Emotion,* 16, 115
Homer: *The Iliad,* 29, 90

Husain, M. F.: *Meenaxi*, 111

"I," 10, 14, 47, 85
Iacoboni, Marco, 77
identity, 2, 14, 86; categorial, 27, 105; continuity and discontinuity, 9–13, 44; practical, 5, 104, 105
ideology, 176, 179, 182; worldview, 107–13
imitations, 1, 62, 137, 147, 159
Indo-European languages, 35, 36
induction, 7, 175
inferential theory, 15, 24, 138, 166
Ingram, John, 42, 46, 76
intentionality, 4, 7,15, 114
interior dialogue, 81
interior monologue, 59, 60, 64, 81, 144
interior movement, 63
interior speech, 59
interior state of mind, 24, 30
interior verbal thought, 70
intermental thought, 66
internalism: and Chomsky, 52–58; literature and dialogue, 58–66
interpretation, 41, 43, 50, 76, 79, 81, 103, 104, 108, 110, 112, 113, 116, 122, 144, 149–54, 157, 158, 150, 161, 162, 163, 164, 165, 166, 170, 171, 172, 176, 179, 184
Iron Man Marathon, 151, 152

Jakobson, Roman, 36
James, William, 74
Jefferson, Thomas, 5
Joyce, James, 62; *Dubliners*, 142; *Finnegans Wake*, 41; *Portrait of the Artist as a Young Man*, 142; *Ulysses*, 60, 61, 142

Kandel, Eric, 121
Kant, Immanuel, 29, 120, 127, 128, 129, 137; *Critique of Judgement*, 133
Kashmir, 21
Kawabata, Hideaki, 120
Keneally, Thomas: *Three Famines: Starvation and Politics*, 167

Kiparsky, Paul, 38
Kringelbach, Morten, 22, 120, 124, 185

Labov, William, 39, 42–43
Lacan, Jacques, 50
Lakoff, George, 73, 74; *More than Cool Reason*, 54
language of thought theory, 44, 56
language science and verbal art, 32–78
LeDoux, Joseph, 2, 3
Left 4 Dead 2, 17
Leonard, Elmore: *The Complete Western Stories of Elmore Leonard*, 118
level of abstraction, 72, 91, 161
leval of generality, 36, 91, 93, 154
Lévi-Strauss, Claude, 51, 173; *Mythologiques*, 49
Li, Ch'ing-chao, 101
Life, 33
limbic system, 17, 116, 166
linguistics, 7, 32–78, 91; past uses in literary study, 34–38
literature: and internalism and dialogue, 58–66; and social betterment, 171–81; and the scope of linguistic theory, 38–52
Llosa, Mario Vargas: *La ciudad y los perros* (*Time of the Hero*); *Los jefes* (*The Cubs and Other Stories*), 109
Lukács, Georg, 111

Mao, Tse-tung, 167
Malouf, David: *Ransom*, 90
Martindale, Colin, 124
Marx, Karl: *Communist Manifesto*, 155
Marxism, 6, 111
Mason, Zachery: *The Lost Books of the Odyssey*, 90
materialism, 72, 74; emergent, 3
Maturana, Humberto, 73
McHale, Brian: "Beginning to Think about Narrative in Poetry," 102
"me," 10, 14
meaning-conferring mechanism, 116, 117
memory, 20, 22, 44, 61, 62, 67, 72, 95, 116,

159, 175; autobiographical, 14; continuity, 11; emotional, 129, 168; long-term, 3; neurosciences, 28; semantic, 76; single, 80; source, 156; working, 11, 12, 12n5, 103
Mexican Revolution, 34, 167
mind: and matter, 4; Vedāntic tradition, 1
"Minimalist Program," 55
mirroring responses, 16, 72
mirror neuron system, 15–16, 77
misattribution, 73, 77–78
Mishra, Rajan, 132
Mishra, Sajan, 132
mood, 119; congruent processing, 107; repair, 20, 22
Moore, Kathleen, 124
morals, 28, 29–30, 33, 104, 127–29, 141, 157, 159, 164, 172, 174, 175, 176, 180, 182; fictional, 113–14; origins, 126
morphology, 38, 60
Mounier, Emmanuel, 9
Mṛcchakaṭikam, 6
myths, 17, 36, 91; Hemingway, 33–34; Lévi-Strauss, 49, 51

Nabokov, Vladimir: *Lolita*, 142
Naccache, Lionel, 3
Nadal, Marcos, 121
NAFTA, 168
narrative fiction, 79–115; character, 94–99; contextualism, 85–88; definitions, 83–85; discourse, selection, and segmentation, 99–104; ethics and emotion, 113–15; modular, 43, 44–45; political and ethical implications, 26–31; and politics, 105–7; processes of simulation, 22, 80; purposes, 154–56; style, 88–94; worldview and ideology, 107–13
neural networks, 3, 67, 70, 82
New Grammarians, 35, 36, 52
New Historicism, 79, 82
Noel, Urayoán: *Hi-Density Politics*, 117
Nolan, Christopher, 13
North American Free Trade Agreement, 168

Oatley, Keith, 80
observer, 4, 47; passive, 48
Ortony, Andrew, 73, 106

Paine, Thomas, 5
Palmer, Alan, 66
Palmer, Linda, 137
parallel distributed processing (PDP), 67, 70–71
parameters, 7, 92, 101
Paris, Jean, 39
Paris Commune, 183
Paris Review, 109
Passos, John Dos, 90, 112; *Brazil on the Move*, 109; *Century's Ebb*, 109
Patañjali: *Yoga Sūtras*, 2
perception, 5, 60, 62, 73, 76, 116–17, 118, 141
peripersonal space, 18
Phelan, James, 54, 56
phrasing, 71, 80, 89, 119
Piera, Carlos, 38
Pinker, Steven, 123, 172
Plato, 57, 146, 147
plots, 145, 168, 180; interest, 154, 157, 160
Poe, Edgar Allan, 84, 90
poetry, 36, 41, 90, 92, 102, 117, 118, 119, 125, 126, 130, 170; sound pattern, 40
political correctness, 86, 87
politics: criticism, 180, 181, 185; narrative fiction, 105–7; and narrative fiction, 26–31; and motivation, 182–86
practical identity, 5, 104, 105
pragmatics, 40, 41, 56
Premchand, Munshi: *Godān*, 92–93
Prentice, Deborah, 81, 156, 178
Prince, Gerald, 54, 83, 81
probabilistic thinking, 20, 21, 23, 33, 96; Bayesian, 103
productivity, 86, 103
prototypes, 120, 122–24, 125, 128, 137, 141
Proust, Marcel, 121; *A la recherche du temps perdu*, 80

quantum mechanics, 4, 99, 146

Ramachandran, V. S., 121; *The Tell-Tale Brain*, 125–26
reasoning, 7, 23, 33, 166; balanced, 19; causal, 96, 164; counterfactual, 45; effective, 185; and emotion, 18, 19
recursion, 50
recursive embedding, 50, 64
reductionism, 3, 5
Reisch, George, 114
religion, 103, 105, 146
Resident Evil, 17
Rizzolatti, Giacomo, 77
Robel, Leon, 39
Rodriguez, Robert, 98, 99, 141, 144, 145, 151, 152; *Machete*, 141; *Planet Terror*, 141; *Roadracers*, 94; *Spy Kids*, 156; *Spy Kids 3*, 148
Ronat, Mitsou: *Language and Responsibility*, 39
Roubaud, Jacques, 39
Russell, Bertrand, 185
Russell, Richard, 123
Russian formalists, 36, 78, 90, 120, 136
Russom, Geoffrey, 38

Said, Edward 132
Sāṁkhya, 1
Sanskrit, 6, 23, 36, 50, 92, 119, 134, 140
Sapir, Edward, 35, 37
Sappho, 120
Saussure, Ferdinand de, 35, 52, 54; *Cours de linguistique générale* (*Course in General Linguistics*), 36, 53
Saussurean linguistics, 35, 52; signifiers, 170
Saw I, 156
Schacter, Daniel, 156
Scheff, Thomas, 81
Scientific American, 83, 90
segmentivity, 60, 102, 118
self: coherent, 12; Patañjali, 4; Vedāntic tradition, 1
selfhood, 9; biology of, 3
self-identity, 11, 14
senses, 6, 34, 72, 77, 117, 135–36, 153
Setti, Ricardo Augusto, 109
Shakespeare, William, 25, 82, 163

simulation, 8, 22–23, 81, 100–101, 112, 115, 138, 139, 142, 144, 184; aversive outcome, 14, 22; Balzac's, 111; cognitive process, 172, 175; counterfactual, 101; effortful, 27; embedded, 64; and emotional response, 60; fictional, 62; hypothetical, 101, 175; Kant's, 129; processes, 80; purpose, 154; tacit, 63; and theory of mind, 13–16, 25, 32
shape, 23, 57, 65, 74, 83, 90, 93, 94, 95–96, 97–98, 100, 102, 103, 116–19, 121, 122, 142, 143, 150, 152–53, 158–59, 160, 166
Sharvit, Yael, 39
Shkovsky, Victor, 136, 147
Skov, Martin, 120
Sollers, Philippe, 39
somatic markers, 72, 73
Sorrows of Young Werther, 174
sound patterns, 40, 41, 89, 92, 130
Sperber, Dan, 40
Spielberg, Steven: *Duel*, 48
Stalinist bureaucracy, 37, 170
Stapp, Henry, 4
statistical tool, 69
Steen, Gerard, 75, 76
Sterne, Laurence: *Tristram Shandy*, 100
story, 37, 47, 48, 50, 83–84, 85, 89, 91, 92, 115, 142, 143, 155, 158, 159, 164, 174; analysis, 94; character, 94–99; development, 95; discourse, 78, 103, 113; duality, 51; dynamism, 118; origin, 93; real, 114; shaping, 181; structure, 44, 46; time, 100. *See also* narrative fiction
story emotion, 14, 22, 23, 34, 154, 157, 160
Stowe, Harriet Beecher, 80
Structuralism, 52, 91
style, 88–94

Tagore, Rabindranath: "Exercise-Book," 80; "Kabuliwallah," 178–79
Tel Quel, 39
thematic roles, 46–48, 49, 50
Theory of Argumentation, 40
Todorov, Tzvetan, 37
Tolstoy, Leo, 34; *Anna Karenina*, 93, 94
tool, analytical. *See* analytical tool

tools, 103, 126, 155, 174; of communication, 65; shape-giving, 158–59; words, 75
transcendental ego, 4
transformational grammar, 37, 49
Trubetzkoy, Nikolai Sergeyevich, 36
truth, 3, 51, 87, 89, 116, 154, 184, 185
Turner, Mark, 39, 106; *More than Cool Reason*, 54
Tversky, Amos, 106

Uncle Tom's Cabin, 129
uniqueness, 12–13, 37, 82, 172

validity, 25, 42, 86, 87, 116
The Valley of the Dolls, 137
Varela, Francisco, 73
Varo, Remedios, 141
Vedānta, 1, 11; Advaita, 2; schools, 2
verbal art: Chomsky and internalism, 52–58; cognitive linguistics and embodied cognition, 72–78; connectionism, 66–72; and language science, 32–78; literature and the scope of linguistic theory, 38–52; literature, internalism, and dialogue, 58–66; past uses of linguistics in literary study, 34–38
Vermeule, Blakey, 94
Vinteuil, 121
virtue, 28, 29, 142, 175, 184
voluntarism, 6, 111
Vuust, Peter, 22, 120, 124, 185
Vygotsky, L. S., 62, 64, 65

"will to style," 89, 97, 109, 115, 155
Wilson, Deirdre, 40
Wilson, Stephen, 77
Woloch, Alex: *The One vs. The Many*, 26
Woolf, Virginia, 62, 81, 143, 155
worldview, 54, 74, 103, 104, 145, 155, 178, 179, 180; and ideology, 107–13, 176
World War II, 180; post-, 35, 37, 153
Wurtz, Robert, 121

Yoga, 2, 4

Zeki, Semir, 120, 122
Zunshine, Lisa, 24, 162n5

www.ingramcontent.com/pod-product-compliance
Lightning Source LLC
Chambersburg PA
CBHW021757230426
43669CB00006B/110